Differentiation Patterns in Higher Plants

Differentiation Patterns in Higher Plants

edited by

KRYSTYNA M. URBANSKA
Geobotanical Department
Swiss Federal Institute of Technology
Zurich, Switzerland

1987

ACADEMIC PRESS
Harcourt Brace Jovanovich, Publishers
London • Orlando • San Diego • New York
Austin • Boston • Sydney • Tokyo • Toronto

ACADEMIC PRESS INC. (LONDON) LTD.
24-28 Oval Road, London NW1 7DX

United States Edition published by
ACADEMIC PRESS, INC.
Orlando, Florida 32887

British Library Cataloguing in Publication Data
Differentiation patterns in higher plants.
1. Plant cell differentiation
I. Urbanska, K.M.
581.1′7 QK725

ISBN 0-12-709425-3

Phototypeset by
Geometra Phototypesetting, Chichester, West Sussex

Printed in Great Britain
at St Edmundsbury Press, Bury St Edmunds, Suffolk

Contributors

Fakhri A. Bazzaz: Department of Organismic and Evolutionary Biology, The Biological Laboratories, Harvard University, 16 Divinity Ave., Cambridge, MA 02138, USA

Christopher D. K. Cook: Institut für Systematische Botanik der Universität Zürich, Zollikerstrasse 107, CH-8008 Zürich, Switzerland

Daniel J. Crawford: Department of Botany, The Ohio State University, Columbus, OH 43210-1293, USA

Alexis Ducousso: Université des Sciences et Techniques de Lille, Laboratoire de Génétique Ecologique et de Biologie des Populations Végétales, F-59655 Villeneuve d'Ascq Cedex, France

Friedrich Ehrendorfer: Institute of Botany, University of Vienna, Rennweg 14, A-1030 Vienna, Austria

William F. Grant: Department of Plant Science, P.O. Box 4000, MacDonald College of McGill University, Ste Anne de Bellevue, Quebec, Canada H9X 1C0

James L. Hamrick: Departments of Botany and Genetics, University of Georgia, Athens, GA 30602, USA

Akira Hiratsuka: Biological Institute, Faculty of Science, Tokohu University, Sendai, Japan

Schoichi Kawano: Department of Botany, Faculty of Science, Kyoto University, Kyoto 606, Japan

R. Bruce Knox: Plant Cell Biology Research Centre, School of Botany, University of Melbourne, Parkville, Victoria 3052, Australia

Elias Landolt: Geobotanisches Institut ETHZ, Stiftung Rübel, Zürichbergstrasse 38, CH-8044 Zürich, Switzerland

David F. Murray: University of Alaska Museum, Fairbanks, Alaska 99701, USA

Shuichi Nakayama: Department of Botany, Faculty of Science, Kyoto University, Kyoto 606, Japan

Daniel Petit: Université des Sciences et Techniques de Lille, Laboratoire de Génétique Ecologique et de Biologie des Populations Végétales, F-59655 Villeneuve d'Ascq Cedex, France

James A. Quinn: Department of Biological Sciences, Rutgers University, Piscataway, NJ 08854, USA

Clive A. Stace: Department of Botany, University of Leicester, University Road, Leicester LE1 7RH, UK

Tod F. Stuessy: Department of Botany, The Ohio State University, Columbus, OH 43210-1293, USA

Sonia E. Sultan: Department of Organismic and Evolutionary Biology, The Biological Laboratories, Harvard University, 16 Divinity Ave., Cambridge, MA 02138, USA

Takenori Takada: Department of Biophysics, Faculty of Science, Kyoto University, Kyoto 606, Japan
Krystyna M. Urbanska: Geobotanical Department, Swiss Federal Institute of Technology, Zürichbergstrasse 38, CH-8044 Zürich, Switzerland
Myriam Valero: Université des Sciences et Techniques de Lille, Laboratoire de Génétique Ecologique et de Biologie des Populations Végétales, F-59655 Villeneuve d'Ascq Cedex, France
Philippe Vernet: Université des Sciences et Techniques de Lille, Laboratoire de Génétique Ecologique et de Biologie des Populations Végétales, F-59655 Villeneuve d'Ascq Cedex, France
Richard Witkus: Department of Botany, The Ohio State University, Columbus, OH 43210-1293, USA

Foreword

The contemporary vigour of the plant biosystematics is amply illustrated by the fine collection of chapters that constitute the present volume. Despite premature accounts of its demise, the detailed experimental study of plant populations continues to attract outstanding scientists. Although the main preoccupation of biosystematics has shifted over the past 50 years from the properties of species and populations to features of the local breeding populations, deductions at all levels of biological organization, when based on experimental studies, continue to be made usefully. The steady growth of the International Organization of Plant Biosystematists (IOPB) itself provides but one measure of the vitality of the field, and the organization of this excellent symposium in Zurich by Krystyna Urbanska constitutes a second such measure.

Traditional biosystematic fields, such as the role of hybridization in plant evolution, the nature of the ecotype, the radiation of species on islands, and the evolution of weedy and colonizing species, are well represented in this book and receive a great deal of attention from biosystematists. On the other hand, new tools such as the comparative analysis of nucleic acids both in the nucleus and in mitochondria and chloroplasts and the investigation of isozymes, are producing new and exciting results. The continued application of such tools, and of powerful mathematical methods of analysis, spurred by the application of computers in a way unthinkable in the 1950s, will produce valuable results for many years to come.

The exploration of breeding systems in plants continues to be of great interest to plant biosystematists. In this connection, C. D. K. Cook's suggestion in this volume that the lack of sexual reproduction may constitute a price to be paid for ecological specialization is an interesting one, and consistent with much of our knowledge. The application of the methods of demography to the understanding of the population structure likewise offers great promise for increasing our understanding of plant evolution. Such an approach is illustrated in this book by the distinguished contribution of Schoichi Kawano and his colleagues.

We are coming to understand all of the facets of plant growth and development better, and this understanding will have important applications in the more completed research of plant evolution. For example, the studies reported by Bruce Knox in this volume, like so much of his excellent work, are most informative about the function of pollen grains and stigmas and thus of incompatibility and the possibility of obtaining hybrids. The understanding of these processes will ultimately help us to assess the evolution of barriers to interspecific hybridization and consequently the functioning of plant species in nature.

In the present volume, about half of the chapters are devoted to what might be called the field of plant population biology. The vigorous and mutually beneficial hybridization of this field with traditional plant biosystematics is nicely illustrated. It will be exciting to study the fruits of this hybrid in more detail at the next IOPB symposium in Japan in 1989.

Peter H. Raven
Missouri Botanical Garden,
St Louis, MO, U.S.A.

Preface

This volume is based on Invited Papers presented at the IOPB Symposium held in Zurich, Switzerland, 13–18 July, 1986. The Symposium was sponsored by the International Organization of Plant Biosystematists, the International Union of Biological Sciences, Swiss Academy of Science, and the Swiss Federal Institute of Technology, Zurich. The generous financial support of the two latter institutions is gratefully acknowledged.

We believe that a better knowledge of differentiation occurring over a wide range of scales and systems of biological organization will result in a more integrated approach to the problem of evolution in higher plants. The purpose of the Symposium was therefore to promote exchange of ideas and experiences between scientists representing various backgrounds and interests. This book is aimed at workers interested in the interface of plant ecology with genetics, demography, reproductive behaviour, etc. May the reader find the chapters that follow as stimulating as the Symposium participants have found the Invited Papers presented.

It is a pleasure to thank various staff members of the Geobotanical Institute, SFIT Zurich, and the Botanical Institute, University of Zurich, who provided assistance during the Symposium. I am also particularly grateful to Ms Susy Dreyer, Ms Anita Hegi, Ms Erica Wohlmann, Gaby Elmer, Marc Nicholls and Martin Schütz, who greatly helped with the preparation of this volume.

K. M. Urbanska
Zurich, Switzerland

Contents

Part III Differentiation in Diverse Biological Systems

Part I

INTRODUCTION

Introductory Remarks

KRYSTYNA M. URBANSKA

Geobotanical Department, Swiss Federal Institute of Technology, Zürich, Switzerland

The complex processes and patterns of differentiation in higher plants operate at *various levels*, from the molecular level of mutation to the biological level of adaptation (Grant, 1985). No particular level can be singled out as the most important.

Since evolution takes place in a continuum of populations through time, all evolutionary processes must, in the final analysis, take place at the population level (Sokal, 1978). The increased interest in plant population biology during the last decade (see e.g. Harper, 1977; Dirzo and Sarukhan, 1984; White, 1985) is undoubtedly influenced by this understanding. The question arises, however, whether the principles of modern population biology are sufficient to explain all evolutionary processes. Differentiation obviously occurs in *diverse biological systems*, and may thus be influenced by some system-specific properties, even if the principal biological rules governing plant life are the same everywhere. Spatial and temporal patterns of differentiation occurring within and among populations, communities, and ecosystems may thus relate to conditions particular to a given system, with historical factors representing an important element.

At the time being, the question stated above remains open, with more new information being required. Another problem is that some data on various systems are available, but remain scattered in the literature and have yet to be brought together.

The knowledge of *diverse aspects* of differentiation is also essential for an accurate assessment of evolutionary potential of a given development. Recent studies in this subject demonstrate that some long-standing beliefs need to be

Differentiation Patterns in Higher Plants
ISBN: 0-12-709425-3

revised. For example, asexual reproduction does not preclude differentiation in some plants (see e.g. Urbanska, 1985; Landolt, 1986). Not only plant reproduction, but also other important aspects are being given increased attention and new insights are being obtained; among them are gene flow (see e.g. Loveless and Hamrick, 1984), phenotypic plasticity (see e.g. Bradshaw, 1974; Schlichting, 1986), or the niche problem (see e.g. Bazzaz, 1986). All these studies show that there is no clear separation between genetics, ecology, and studies in plant behaviour, and that approaches from diverse fields of plant life science urgently need to be integrated.

Our purpose in organizing this volume was to treat differentiation in higher plants in its diversity, the contributors having been selected accordingly. This book does not cover all possible aspects of differentiation as no single volume can do this; rather, it was intended to present some ideas and examples that might prove useful for future research. The first part of the book focuses on diverse levels and aspects of differentiation. The first two chapters are concerned with the genetic component of plant evolution; they discuss genome differentiation (Grant, Chapter 1), and genome transcription patterns in differentiation of gametes (Knox, Chapter 2). The subsequent chapters develop the theme of interplay between genetic and ecological factors for particular aspects of differentiation: gene flow (Hamrick, Chapter 3), variation in environmental parameters (Bazzaz and Sultan, Chapter 4), and phenotypic plasticity (Quinn, Chapter 5). The last chapter of Part II deals with natural hybridization, which does not merely represent a re-fusion of differentiated genomes, but often leads to a further differentiation (Stace, Chapter 6).

In Part III examples of differentiation in diverse biological systems are presented. Various patterns are shown in adjacent populations of a single species inhabiting two different environments (Vernet *et al.*, Chapter 7), different species of the same family occurring in temperate woodland within a rather limited area (Kawano *et al.*, Chapter 8), congeners on oceanic islands (Crawford *et al.*, Chapter 9), and also in a single family of aquatic plants distributed all over the world (Landolt, Chapter 10). Species representing various families are discussed in chapters concerned with aquatic weeds (Cook, Chapter 11), and tropical woody plants (Ehrendorfer, Chapter 12). Finally, the whole vascular flora of a large Arctic region is dealt with (Murray, Chapter 13).

So far as I am aware, most of the material presented here is new and has not previously appeared in book form.

References

Bazzaz, F. A. (1986). Experimental studies on the evolution of niche in successional plant populations: a synthesis. *Sp. Publ. Ser. Br. Ecol. Soc.* **26**, 245–272.

Bradshaw, A. D. (1974). Environment and plasticity. *Brookhaven Symp. Biol.* **25**, 75–94.

Dirzo, R. and Sarukhan, J. (eds) (1984). "Perspectives in Plant Population Ecology". Sinauer Associates Inc., Sunderland, MA.

Grant, V. (1985). "The Evolutionary Processes: A Critic Review of Evolutionary Theory". Columbia University Press, New York.

Harper, J. L. (1977). "Population Biology of Plants". Academic Press, London.

Landolt, E. (1986). The family of Lemnaceae — a monographic study. *Veröff. Geobot. Inst. ETH Stiftung Rübel*, **71**.

Loveless, M. D. and Hamrick, J. L. (1984). Ecological determinants of genetic structure in plant populations. *Ann. Rev. Ecol. Syst.* **15**, 65–95.

Schlichting, C. D. (1986). The evolution of phenotypic plasticity in plants. *Ann. Rev. Ecol. Syst.* **17**, 667–693.

Sokal, R. R. (1978). Population differentiation: something new or more of the same? *In* "Ecological Genetics: the Interface" (P. F. Brussard, ed.), pp. 215–239. Springer Verlag, New York, Heidelberg.

Urbanska, K. M. (1985). Some life history strategies and population structure in asexually reproducing plants. *Bot. Helv.* **95**, 81–97.

White, J. (1985). "Studies in Plant Demography". Academic Press, London.

Part II

DIFFERENTIATION LEVELS

1 Genome Differentiation in Higher Plants

WILLIAM F. GRANT

Department of Plant Science, Macdonald College of McGill University, St Anne de Bellevue, Quebec, Canada

I. Introduction

In higher organisms, two or more genomes derived from parental gametes, with their content of hereditary information, form the eukaryotic chromosome complement or karyotype. On the molecular level, the genome can be defined as the complement of all the DNA sequences found in the chromosomes (Flavell, 1983). The arrangement of the sequences with respect to one another determines the architecture of the genome. As a result, the genome is a major component of a system which dictates growth and reproduction, heredity and evolution. Everything that happens in the genome is related to what happens in the organism, or in retrospect, to what has happened in its ancestors. Genome evolution can be attributed to changes in structure, organization, sequence composition and size of chromosomes (Flavell, 1982), which result in the breakdown of chromosome homology, and ultimately in genetic isolation.

Karyological studies have played a prominent role in biosystematic research for many years, and it may be expected that they will continue to do so, as has been shown by the many advances that have been made in the study of the karyotype in the past decade (Grant, 1984). The proliferation of recent articles and texts concerning the genome and its relationship with systematics indicates the continued importance of this discipline in our understanding of phylogenetic relationships (Bennett, 1984; Brandham and Bennett, 1983; Cavalier-Smith, 1985a; Dutta, 1986; Ehrendorfer, 1983; Grant, 1984; Sharma and Sharma, 1985).

Differentiation Patterns in Higher Plants
ISBN: 0-12-709425-3

The word "Genome" has become the new title for the *Canadian Journal of Genetics and Cytology*, indicating the many levels of discernment that this word implies.

This chapter will review various aspects of the genome and its role in systematics, stressing information derived from cytogenetical and molecular techniques (Table 1). Specifically, discussions will be concerned with various aspects of chromosome numbers and polyploidy, chromosome morphology (banding), nuclear DNA content, and the more recent use of molecular approaches to systematics.

TABLE 1
Genomic characters and biosystematics (modified from Bennett, 1984)

Cytogenetical	Molecular
1. Chromosome number (a) Gametic no., *n* (b) Genomic no., *x* (c) Polyploidy: auto-, allopolyploidy dysploidy	1. Amount of DNA per genome or chromosome 2. Base composition: amounts of unique and repeated sequences in the genome; coding and noncoding sequences; rDNA variation
2. Karyotype (a) Intranuclear spatial order of chromosomes (b) Karyotype repatterning (c) Chromosome repatterning (d) B chromosomes	3. DNA–DNA, DNA–RNA hybridization 4. Restriction maps of DNA sequences and organelle genomes
3. Chromosome morphology (a) Heterochromatin: chromomeres, chromocenters, telomeres, knobs (b) Centromeres (c) NORs, satellites (d) Banding patterns	
4. Chromosome pairing behaviour (a) Chiasma frequency (b) Asynapsis-Desynapsis	

II. Chromosome Studies

A. Chromosome numbers

Chromosomes constitute special information, as they (a) encode the genetic information which results in the phenotypic expression of the characters, and (b) participate in recombination which gives rise to the variability between populations.

However, our knowledge of chromosome numbers *per se* for the world's flora is still far from complete and may never be complete with the rapid spread of "civilization" (Wilson, 1985). Since Stace's (1980) tabulation of known chromosome numbers for the world's flora (6% for bryophytes, 20% for pteridophytes and only 15–20% of the angiosperms), the *World Wildlife Fund International Plants Conservation Programme* has projected a loss of 1 in 6 species of flowering plants before the middle of the next century. Even the chromosome numbers of many close relatives of our cultivated species are still to be determined (Hsiao *et al.*, 1986).

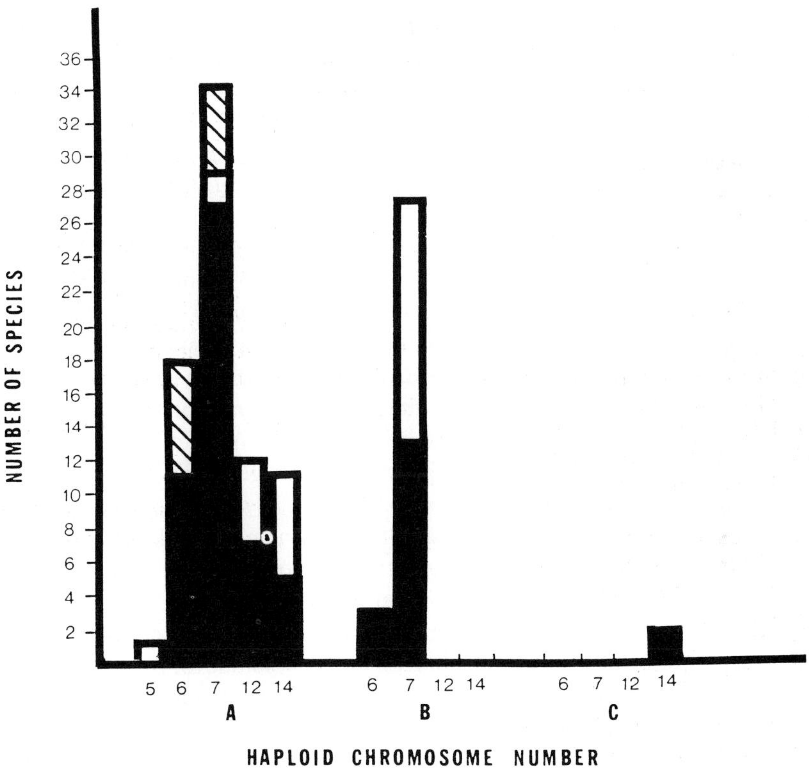

FIG. 1 Geographic distribution of *Lotus* species in relation to chromosome number. A, Eur-Afro-Asian species; B, North American species; C, Australian species. Total of A, B and C represent the species of *Lotus* for which chromosome numbers are known. Cross-hatching represents 12 taxa with both diploid and tetraploid cytotypes; clear areas represent chromosome numbers for taxa since 1965 including one species with a basic chromosome number of 5.

A time frame for knowing all of the chromosome numbers in a genus may be extrapolated from my compilation of chromosome numbers for the genus *Lotus* (Grant, 1986). Since 1965, chromosome numbers for 32 new species of *Lotus* have been reported (Fig. 1). This represents a 29.6% increase in the number of species for which chromosome numbers have been determined since this date, however, at this rate — essentially one new species per year — it will require almost another century for the chromosome numbers to be determined for the remaining species of the genus!

Recent data have shown that information derived from chromosome numbers has not lost its excitement in helping to resolve phylogenetic relationships. In the genus *Acmella* (Asteraceae), Jansen (1985) has reported 372 chromosome counts in 26 taxa and has shown that differences in chromosome numbers have clearly delineated some pairs of closely related taxa. In addition, he has found polyploidy in 16 out of 27 taxa, emphasizing the importance of polyploidy in speciation in the genus.

Continuing karyotype analyses may be expected to uncover (a) additional polyploid taxa, as recently reported for *Agrostis* (*Agr. hideoi*, $2n = 2x = 14$; *Agr. valvata*, $2n = 4x = 28$; *Agr. Scabra*, $2n = 6x = 42$; Tateoka, 1985), and (b) the origins of allopolyploid taxa, for example, *Ranunculus cantoniensis* ($4x$), which has been shown to be an amphidiploid of *R. silerifolius* ($2x$) and *R. chinensis* ($2x$) (Okada, 1984).

That allopolyploids may *repeatedly* arise by hybridization and chromosome doubling in a population is reported by Werth *et al.* (1985) in the genus *Asplenium* for the allotetraploids *Asp. bradleyi* and *Asp. pinnatifidum*. These allotetraploids are considered to have arisen more than once as minor genetic differences were found between individuals by allozyme analyses. The pattern of allelic variation in the allotetraploids is considered to be the result of repeated allopolyploidization involving diploid pairs of different genotypes and not the result of mutation following hybridization. The authors ask the question, "How general is the phenomenon of recurrent origins of allopolyploid species?" They cite studies in *Triticum, Phlox* and *Asp. lepidum* which have shown that differently named species may represent analogous combinations of different morphotypes of the same diploid species. It is expected that recurrent origins of other allopolyploid taxa will be found by allozyme analyses. The phylogenetic significance of such a process will have to be assessed. What is the rate of speciation on the polyploid level from the repeated production of allotetraploids from the same diploid progenitors which possess minor heterozygosity?

With the study of chromosome numbers of new collections, species with dysploid and dibasic chromosome numbers may be found (Zinov'eva-Stahevitch and Grant, 1984a). In the genus *Impatiens*, where considerable polysomaty occurs, great care must be taken when reporting chromosome numbers to distinguish between dysploidy and polysomaty (Zinov'eva-Stahevitch and Grant, 1984b). For example, a plant of *Impatiens auricoma* Baill. was recently studied and cells with $2n = 8$ were found. Further examination showed that the majority of cells had $2n = 16$.

While the determination of chromosome numbers *per se* lacks the glamour and funding of "molecular genetics", it may be readily seen that continued support is needed for the determination of chromosome numbers, especially from tropical regions (Iltis, 1983). Again, this information is vital for our understanding of the phylogenetic relationships of plant populations, especially when one considers that 70% of all higher plants are estimated to be polyploids (Gottschalk, 1985).

Some help may be expected from the use of digitizers and microcomputers. These aids will increase the speed and accuracy of chromosomal measurements, as well as providing a rapid means for the illustration of idiograms (Green *et al.*, 1984; Fukui, 1986). Their use will steadily increase.

B. Chromosome markers

As the number of chromosomes in a complement increase, it becomes more difficult to differentiate between individual chromosomes. However, some pairs of chromosomes in a karyotype may be identified from the others and may be used as markers. Such is the case in *Impatiens* where there exists a small wholly pycnotic dumbbell-shaped chromosome. This chromosome was found in karyotypes of related taxa which represented a single monophyletic group (Zinov'eva-Stahevitch and Grant, 1985). Such a marker chromosome readily separated these species from closely related taxa. Another interesting chromosome, that also served as a marker chromosome, is a submetacentric chromosome found in *Cryptomeria japonica* ($2n = 22$). This chromosome is readily distinguished from the other chromosomes of the complement by its exceptionally long centromeric region (Schlarbaum *et al.*, 1984).

Armstrong (1985) has produced a hybrid between *Bromus altissimus* and *B. arvensis*. He has observed a large heterochromatic telomere in a chromosome of the complement of *B. arvensis* which readily distinguishes the *B. arvensis* genome from that of *B. altissimus* in the somatic karyotype of the hybrid. The presence of B chromosomes may also serve to indicate a hybrid taxon. B chromosomes have occurred in hybrids between diploid species of *Lotus* when *L. alpinus* (2*x*) was one of the parents (Somaroo and Grant, 1971).

C. Genome in hybrids

1. Preferential staining of the whole genome. Armstrong has found that the two *genomes* which compose the karyotype of the hybrid *Bromus altissimus* × *B. arvensis* show differential staining after C banding. The complete *genomes* of the two species can be readily distinguished in the somatic hybrid complement by the difference in staining reaction (Fig. 2). The chromosomes from the male parent (*B. arvensis*) are largely heterochromatinized at Metaphase I. It would be of interest to know the effect of heterochromatinization on the reproductive process, and whether the entire genome may become eliminated as in the mealy bug (Brown, 1966).

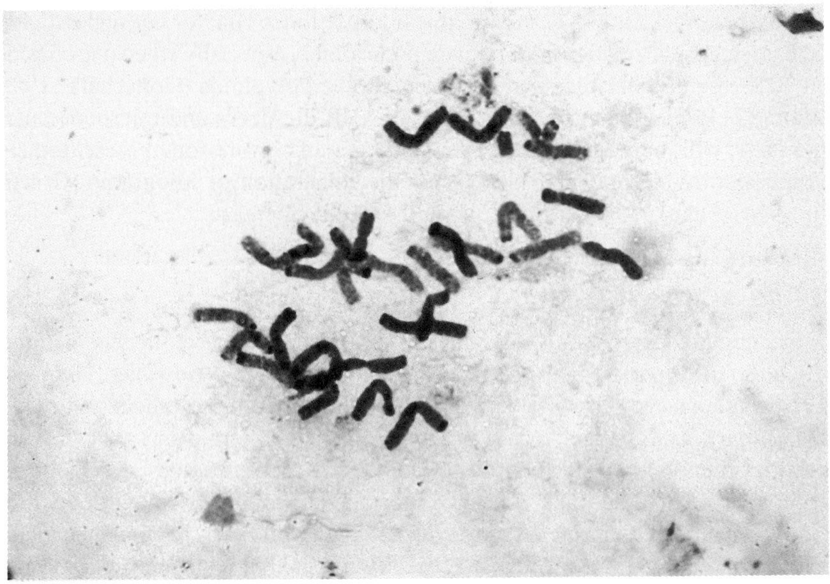

FIG. 2 Somatic chromosomes of *Bromus arvensis* × *B. altissimus* amphidiploid illustrating differential staining of the two genomes (from Armstrong, 1985; by permission of the author and the Genetics Society of Canada).

2. Genome orientation. Interest in the orientation of the chromosomes of parental genomes in a hybrid has led to the idea that the haploid genome is a basic structural unit in nuclear architecture. Finch and colleagues (see Bennett, 1984 for review) have shown that there is a tendency for haploid genomes to be concentrically spatially separated in interspecific diploid hybrids. Further evidence to support this hypothesis has come from studies of Linde-Laursen and Jensen (1984) who have shown a similar separation of parental genomes in hybrids between *Hordeum vulgare* and some polyploid alien species. The "vulgare" and the "non-vulgare" chromosomes, respectively, were located closer to each other than would be expected on a random basis, further supporting the haploid genome as a basic structural unit in nuclear architecture.

Bennett has devised a model for the specific order of chromosomes in the haploid genome in which he has shown that individual chromosomes are arranged in a highly ordered sequence with their neighbours. In general, the arrangement of each chromosome depends on the size of its arms. Changes in DNA content of adjacent pairs of long or short arms of chromosomes can be predicted and shown by C banding. Details regarding methodology and examples have been given in detail (Bennett, 1984). A similar arrangement of the chromosomes in the interphase nucleus of

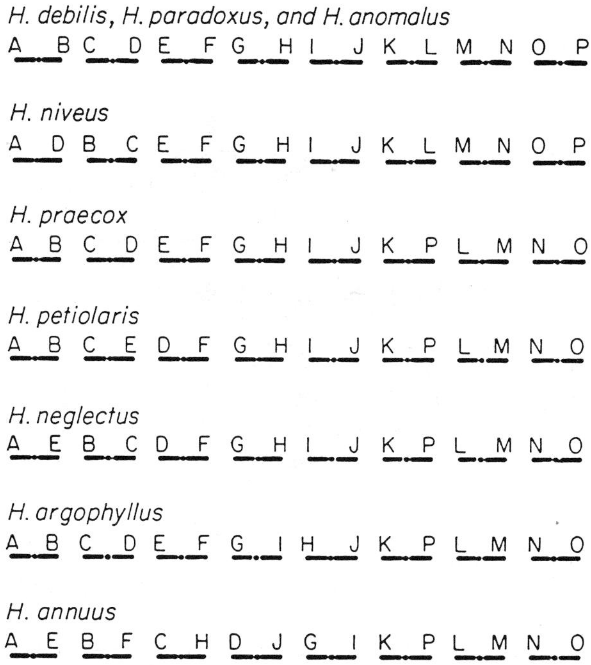

FIG. 3 Chromosome end arrangement in *Helianthus* species. The entire complement of chromosomes for each taxon is not shown (from Chandler *et al.*, 1986; by permission of *Systematic Botany* and the authors).

Secale cereale has been proposed by Lavania and Sharma (1985). They have shown that homologus chromosomes lie adjacent and chromosome arms of similar length are associated at the telomeric ends. Chandler *et al.* (1986) have proposed end arrangements of chromosomes for a number of species of *Helianthus* based on the observed multivalent configurations formed in interspecific hybrids (Fig. 3). The authors postulate that the end arrangements explain most of the hybrid pairing configurations.

III. Nuclear DNA and Systematics

A. DNA content

The mass of DNA in an unreplicated genome, as in an immature pollen grain prior to DNA synthesis, is generally referred to as the C-value. Unreplicated diploid nuclei contain twice that amount of DNA, and therefore, have a C-value of 2. Many characters have been associated with the C-value that affect or control the phenotype of a plant which Bennett (1971) has referred to as the *nucleotype*. These "non-genic" characters of mass or size of the nuclear DNA (genome size) are considered to be independent of its encoded informational content. Some of these characters which correlate with the C-value are listed in Table 2.

TABLE 2
Some characters which correlate positively with C value[a]

1. Total length, volume and mass of the metaphase chromosome (Nagl and Ehrendorfer, 1973; Bennett, 1985; Sims and Price, 1985)
2. Nuclear volume (Baetcke *et al.*, 1967; Edwards and Endrizzi, 1975)
3. Cell volume, cell size (Price *et al.*, 1973; Cavalier-Smith, 1985b,c; Nurse, 1985)
4. Centromere volume (Jenkins and Bennett, 1981)
5. Seed weight (Bennett, 1972; Davies, 1977)
6. Seed development, growth rate (Gustafson and Bennett, 1982; Grime, 1983; Bennett, 1985)
7. Length of cell cycle (Van't Hof, 1965; Evans and Rees, 1971; Bennett, 1972; Nagl, 1974)
8. Duration of meiosis (Bennett, 1985)
9. Rate and duration of DNA synthesis (Van't Hof, 1975)
10. Pollen grain volume at maturity (Karp *et al.*, 1982)
11. Life cycle; minimum generation time (Bennett, 1985; Nagl amd Ehrendorfer, 1974)
12. Climatic factors (temperature, rainfall, water stress) (Bennett, 1976; Grime, 1983; Levin and Funderberg, 1979; Stebbins, 1976; Price *et al.*, 1981a,b)
13. Ecological and phenological factors (Grime and Mowforth, 1982; Martinez and Ginzo, 1985)
14. Latitude and altitude (Bennett, 1985; Price *et al.*, 1981a; Kenton, 1984; Martinez and Ginzo, 1985)
15. Optimum environment and geographical ranges of crop and non-crop species (Bennett, 1985)
16. Length of the synaptonemal complex (Anderson *et al.*, 1985)
17. Spatial chromosome arrangement (Bennett, 1984)
18. Ultraviolet light intensity (Bennett, 1976)
19. Chloroplasts per stomatal guard cell (Butterfass, 1983)
20. Radiosensitivity (Underbrink and Pond, 1976)
21. Forward mutation rate per locus per rad (Abrahamson *et al.*, 1973)
22. Sister chromatid exchange (Cortes *et al.*, 1983)

[a]Additional and earlier references may be found in Bennett (1985) and in the references cited.

An examination of cytophotometric measurements has shown that DNA content varies widely among higher plant species (> 2500-fold) and may even vary considerably between plants within a single species (*Poa annua* (80%), *Zea mays* (37%); Bennett and Smith, 1976; Bennett, 1985). Intraspecific variation of more than 4% is reported for 24 species of angiosperms, 16 of these since 1980 (Bennett, 1985). Continued interest in DNA variation stems from the large and still unexplained differences that exist both on the intrageneric and intraspecific level.

Nevertheless, DNA nuclear measurements have been useful in delimiting taxa.

Sims and Price (1985) have recently shown that there is a difference of 162.0 absorbency units between the species of *Helianthus* with the highest DNA content (*H. agrestis*) and the species with the lowest DNA content (*H. neglectus*), and that the annual species which have a lower DNA value were significantly different from that of the perennial species (means 70.05 compared with 97.70 units).

From a study of the DNA content of 22 lines of *Zea mays*, Rayburn *et al.* (1985) found 23% variation among the varieties, and that both knob number, and the amount of heterochromatin, positively correlated with DNA content. In *Zea mays*, in contrast to that reported for other genera, DNA content is negatively correlated with increasing latitude. Rayburn *et al.* (1985) suggest that selective pressures, via nucleotypic effects for individual plant yield in cooler climates with shorter growing seasons, would account for the negative correlation with latitude. As pointed out by Bennett (1985), the nucleotype limits the range of phenotypic expression, which in the case of a species with a very high C-value prolongs development, and may preclude the establishment of such a species within a growing season in a cold environment, e.g. high latitude.

Additional observations are needed to determine (1) the extent that selection for genome size is followed by changes in phenotypic characters and, (2) how closely selection in nucleotypically determined characters is followed by genome size changes. From studies in triticales, Bennett (1985) has suggested that it may be possible to select a plant with reduced DNA C-value provided it involved a reduction in non-genic heterochromatin. That selection for such a plant is possible comes from the results of a study in which a triticale plant lacking heterochromatin normally present in the telomeres of chromosomes of the rye parent was established in four years. This plant had less telomeric heterochromatin and a DNA C-value 3.5% less than one of the parents (Gustafson and Bennett, 1982). This shows that significant reduction in DNA values can be achieved.

While the DNA C-value in polyploids is generally a multiple of the C-value (Vida and Mohay, 1980), certain exceptions have been noted. In the genus *Betula*, there is a polyploid series ($2x$ to $6x$) in which plants of the highest ploidy have a DNA value considerably less than that expected on a theoretical basis (Table 3).

TABLE 3

Nuclear DNA absorbance for different ploidy levels of *Betula* relative to absorbance of standard ($2n = 28$) given the value of 1.0 (modified from Grant, 1969)

$2n$	Ploidy	DNA absorbance		Average DNA per chromosome
		Observed	Theoretical	
28	$2x$	1.0	1.0	3.57
42	$3x$	1.41	1.5	3.36
56	$4x$	1.96	2.0	3.50
70	$5x$	2.42	2.5	3.46
84	$6x$	2.29	3.0	2.73

The reduced DNA value reflected a reduction in the size of some of the chromosomes or sets of chromosomes in these plants (Taper and Grant, 1973). Probst (1972) found a similar case in *Leucanthemum* where plants of the highest ploidy had a nuclear DNA content less than the sum of the component genome sizes (Table 4).

TABLE 4

Nuclear DNA absorbance for different ploidy levels of *Leucanthemum vulgare* relative to absorbance of the standard ($2n = 18$) given the value of 1.0 (modified from Probst, 1972)

$2n$	Ploidy	DNA absorbance	
		Observed	Theoretical
18	$2x$	1.0	1.0
36	$4x$	1.98	2.0
72	$8x$	2.5	4.0

Murray (1985) has reported a similar case in *Polypodium* in which the polyploids have component genomes smaller than those of the diploids and a relatively smaller total chromosome length than the diploids (Fig. 4). In the *Tradescantia crassifolia* and *T. virginiana* alliances, Martinez and Ginzo (1985) found that the DNA content per genome decreases when the ploidy level increases at each ploidy level (Table 5). Poggio and Hunziker (1986) also report an octoploid species of *Bulnesia* which has a lower DNA content than two diploid species (Fig. 5).

TABLE 5

Nuclear DNA absorbance for different ploidy levels of *Tradescantia brevifolia* and *T. crassifolia* relative to absorbance of the standard ($2n = 12$) given the value of 1.0 (modified from Martinez and Ginzo, 1985)

$2n$	Ploidy	*T. potosina:* DNA absorbance		*T. crassifolia:* DNA absorbance	
		Observed	Expected	Observed	Expected
12	$2x$	1.00	1.0	1.00	1.0
24	$4x$	2.07	2.0	1.81	2.0
36	$8x$	2.41	3.0	2.23	3.0

How do we account for these polyploid species with lower genomic amounts of DNA than their related diploids? It is well known that changes in nuclear DNA in the evolution of species stem from changes in the amounts of both repetitive and non-repetitive DNA sequences (Raina and Narayan, 1984). Bachmann *et al.* (1985) suggest that differences in genome size could arise by a "saltatory" process whereby a molecular mechanism cuts out repetitive sequences of DNA from all chromosomes simultaneously. Polyploids with reduced DNA content would be expected to arise from saltatory changes in genome size occurring on a lower ploidy level, followed by a period of genome size adjustment through directional

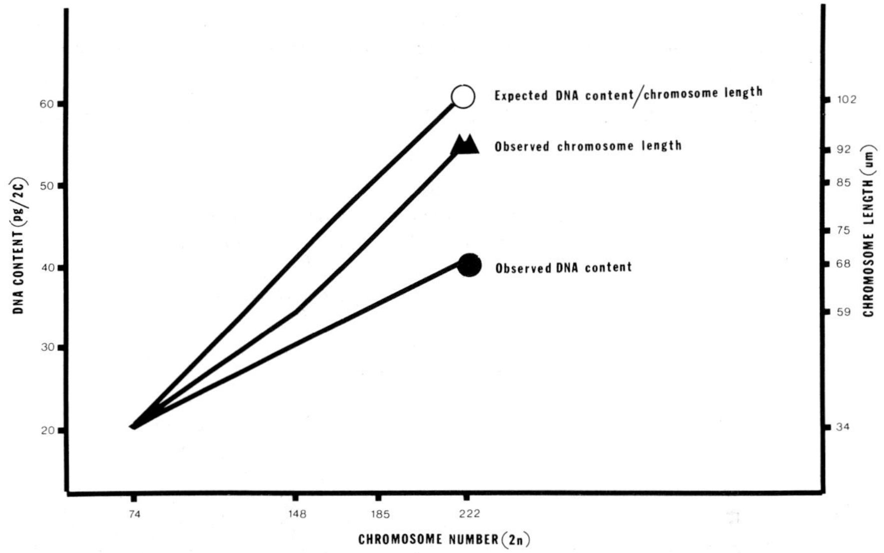

FIG. 4 Chromosome number, chromosome length and DNA amounts in species of *Polypodium* showing reduction in chromosome length and DNA content at high ploidy level (data from Murray, 1985).

selection. That entire chromosomes of a genome can be affected by gains and losses in amounts of DNA has been summarized by Rees (1984), who has shown that with an increase in nuclear DNA amount, the small chromosomes of the genome acquire as much extra DNA as the large chromosomes. Chromosome length *per se* at metaphase is largely dependent upon the duration of the mitotic cycle (Karp *et al.*, 1982) and thus quite independently of their DNA content (Fivawo and Rees, 1985).

B. Chromosome banding

Chromosome banding is an exciting cytological approach which has given to the chromosomes topographical landmarks that help resolve systematic and evolutionary studies on all taxonomic levels (Fukuda, 1984; Vosa, 1985). Vosa (1985) has recently reviewed the history of the various techniques which elicit banding and the various theories on their biochemical induction associated with chromosome structure. Since the 1940 classical cold-induced DNA-starvation method of Darlington and La Cour, a number of banding techniques have become widely used. These include fluorescent dyes, such as quinacrine and ethidium bromide, the fluorochrome Hoechst 33258, C-banding (constitutive heterochromatin), Hy-banding and N- and R-banding. The banding pattern in plants is limited by the fact that plant chromosomes do not show G-bands and that all banding is correlated with heterochromatin distribution within the chromosome. The amount of heterochromatin can vary considerably between species in different sections of a genus, as shown in the studies of *Tulipa* by Blakey and Vosa (1982). Species

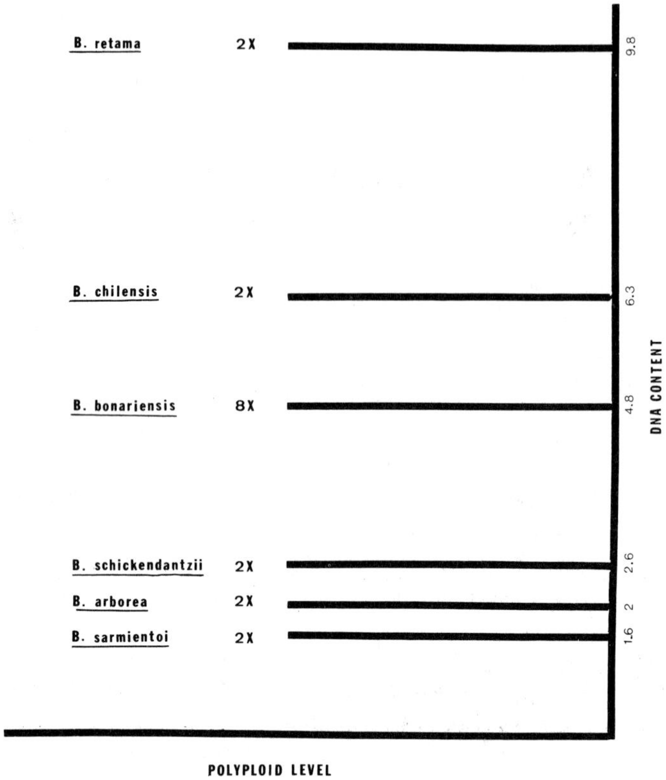

FIG. 5 DNA amount in species of *Bulnesia* in relation to ploidy showing lower DNA content of the octoploid species *B. bonariensis* (data from Poggio and Hunziker, 1986; by permission of the American Genetics Association).

in the subgenus *Eriostemones* have a heterochromatin content up to 15%, whereas in the subgenus *Leiostemones* the heterochromatin content reaches 45%.

Chromatin composition in species of *Pellia* (Hepaticae) is of particular interest, for it was in this liverwort genus that heterochromatin was first identified by Heitz. *Pellia neesiana*, which is a dioecious species, has 28% heterochromatin in the genome of male plants and 35% in the genome of female plants (Newton, 1985). Newton (1985) has recently studied C-, Q-, N- and Hoechst 33258-banding in *P. epiphylla* and *P. neesiana* and has noted marked differences in C-banding pattern between the two species.

It has long been known that the knobs in pachytene chromosomes of *Zea mays* are heterochromatic. Rayburn *et al.* (1985) have C-banded the somatic metaphase chromosomes of 21 lines of *Zea mays* and have shown that the C-banding pattern correlates with the heterochromatin of the pachytene knobs.

Plants possessing small chromosomes are limited by their size in the amount of total chromatin they possess, and therefore in the amount of heterochromatin which they are likely to contain in segments of sufficient size to show banding.

As a result, satisfactory banding in plants possessing small chromosomes may not been achieved. Also, the use of C-banding may not improve the level of identification of taxa beyond that reached by the use of chromosome morphology (Linde-Laursen *et al.*, 1986).

Despite these limitations, chromosome banding has been of considerable value in systematic and cytogenetic studies. For example in the latter field, C-banding has been used to distinguish chromosomes in reconstructed karyotypes of *Vicia faba*, the sex chromosomes in *Rumex*, trisomics of *Secale cereale* and translocations between wheat and barley and wheat and rye (references in Vosa, 1985). Linde-Laursen (1984) has identified by Giemsa C-banding all four homologues in autotetraploid barley ($2n = 28$) and some aneuploids derived from the autotetraploids.

A recent review lists 45 genera of higher plants in which chromosomes have been banded for one or more species (Fukuda, 1984). Species of *Trillium* provide excellent illustrations of banding after cold treatment (Fukuda and Grant, 1980). By this technique, species (Chinnappa and Morton, 1978) and populations of *Trillium* (Fukuda and Grant, 1980) have been clearly differentiated. For example, in an analysis of 25 populations of *Trillium grandiflorum* examined from throughout the natural range of the species in eastern North America, it was found that chromosome "D" varied three to five times more frequently in its banding pattern than chromosomes A, B, C and E. This was in contrast to findings in populations of *T. ovatum* in western North America in which chromosome "C" was most variable, and likewise, to findings in populations of *T. kamtschaticum* in Japan, in which chromosome "A" was most variable (Fukuda, 1984).

IV. Molecular Studies

A. Molecular systematics

Because of polyploidy, higher plants have genomes which range considerably in size from about 2×10^8 to over 10^{11} nucleotide pairs (see list of species, Table I in Sorenson, 1984). Almost all eukaryotes have genomes with far greater DNA complexities than is consistent with the estimated number of genes (5000 to 50 000) present or required to code for the proteins necessary for an organism. According to Cavalier-Smith (1985d), only a small fraction of nuclear DNA makes up the structural and the regulatory genes, while 30% to 99.99% (0.003 pg to over 300 pg, a 10^5-fold range) of the DNA is "redundant" or "silent" and is composed of DNA of various degrees of repetitiveness — from short repeats (satellite DNA), to intermediate repeats and to longer, more complex ones. This apparent excess of DNA in the genome which is "redundant" is usually referred to as the C-value paradox. As will be shown from recent studies on the molecular level, not only is DNA architecture important, but plasmid and genetic recombinant processes are also becoming important for the elucidation of systematic and evolutionary relationships of taxa.

B. Restriction mapping

Speciation and evolutionary divergences are being shown by a number of biochemical procedures, many of which are still in a state of rapid evolution themselves. One of these is restriction mapping in which restriction endonucleases are used to recognize short nucleotide sequences in DNA and cleave the molecule at these specific sites. This procedure makes possible the diagramming of positions of cleavage by the restriction enzymes so that comparisons can be made between taxa on the DNA level. The development of restriction endonuclease digestion, cloned probes, and other recombinant DNA techniques, has allowed direct comparisons of homologous DNA segments among species. The power of these DNA techniques for systematic and evolutionary studies lies in their direct analysis of the genotype, rather than of derived phenotypes, and in their ability to detect numerous mutation events that can be used to infer phylogenies (Sytsma and Schaal, 1985). The majority of the studies using these techniques to resolve phylogenetic problems in plants has been confined to the small genome of chloroplasts (Bowman *et al.*, 1983; Clegg *et al.*, 1984; Palmer *et al.*, 1983; Perl-Treves *et al.*, 1985). Mitochondrial DNA of plants is both more difficult to isolate and more complex than is chloroplast and nuclear DNA (Stern and Palmer, 1984). Cell suspension cultures may be used as the tissue source when difficulty is encountered with the isolation of organelle DNA from leaf material with high phenolic levels (DeBonte *et al.*, 1984). Since nuclear DNA is larger than organelle DNA by many orders of magnitude, restriction fragment analysis is almost impossible to do without cloned probes (radioactive DNA or RNA; Doyle *et al.*, 1984, or immunofluorescent labelling; Trask *et al.*, 1985) that can isolate specific DNA segments. Among the most useful nuclear genes for systematic studies are those encoding the ribosomal genes. These are suitable for use at a number of taxonomic levels because they are mid- to highly-repetitive and contain both a slowly evolving portion (the transcribed 5S, 18S, and 25S rRNA genes) and a rapidly evolving portion (the non-transcribed spacer; Appels and Dvorak, 1982). The majority of studies using restriction mapping have been carried out in cultivated species.

Recently, Furner *et al.* (1986) have shown that the genome of *Nicotiana glauca* contains a large inverted repeat segment of DNA of at least 13 kilobases in length (Fig. 6). This region is derived from a root-inducing plasmid of *Agrobacterium rhizogenes* which provokes neoplastic outgrowths on higher plants by transferring DNA from the root-inducing plasmid to the host genome. This inverted repeat region was found in the genome of uninfected *Nicotiana glauca* plants and in certain species in the subgenera *rustica* and *tabacum*. The species *Nicotiana otophora*, *N. tomentosiformis*, *N. tomentosa* and *N. benavesdesii* all show homology for this inverted region. DNA probes from both *A. rhizogenes* and *N. glauca* hybridized to similar bands in each species. The authors propose a sequence of evolution as follows: the DNA homology results from an

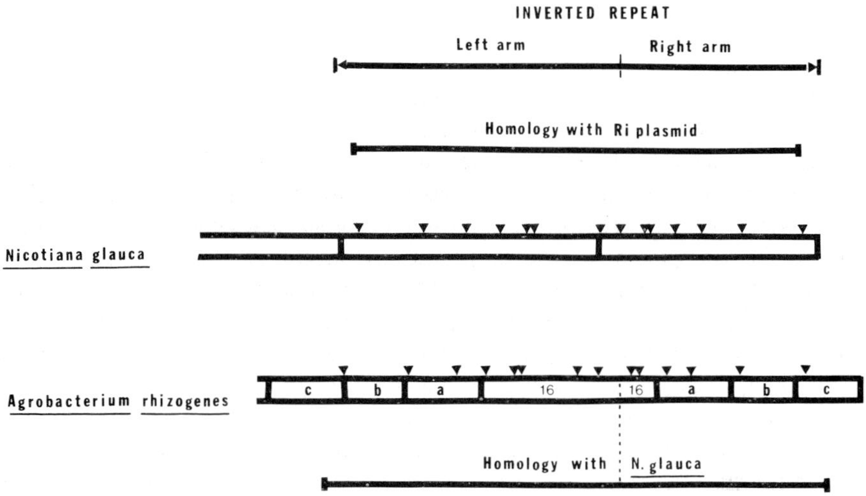

FIG. 6 A restriction map of the Ri (root inducing) plasmid which is an imperfect repeat indicated by the dotted line. The genome of *N. glauca* contains the inverted repeat which is homologus with the DNA of *A. rhizogenes*. Common restriction sites are indicated by arrows (data from Furner *et al.*, 1986; reprinted by permission of the author and *Nature*, copyright 1986 Macmillan Journals Limited).

Agrobacterium infection early in the evolution of the genus *Nicotiana*, resulting in the incorporation of this DNA into the *Nicotiana* genome. This gave rise to transformed fertile progeny. Subsequent evolution of the transformed plants gave rise to the subgenera *rustica* and *tabacum*. The data suggest that this infection occurred in the species complex preceding the divergence of pre-*tabacum* and pre-*rustica* lines. Species lacking detectable homology may result from subsequent loss of these DNA sequences. An alternative possibility is a late origin and lateral spread throughout these two subgenera by interspecific hybridization.

Doyle *et al.* (1985) have obtained evidence from restriction enzyme mapping of the 5S and 18S – 25S rDNA genes to confirm evidence of intergeneric hybridization between *Tolmiea menziesii* and *Tellima grandiflora* by determining the repeated lengths of the ribosomal genes. The 5S ribosomal gene repeat length is approximately 480 and 450 base pairs, respectively, in *Tolmiea* and *Tellima*. The repeat length of the 18S – 25S genes varied from 11 – 13 kb in *Tolmiea*, and was only about 9 kb in *Tellima*. The putative hybrid combined the repeat lengths of both *Tolmiea* and *Tellima*, confirming its hybrid nature.

Sytsma and Schaal (1985) have used DNA restriction mapping to study the variation in nuclear ribosomal DNA (rDNA) and chloroplast DNA (cpDNA) to aid in a phylogenetic reconstruction of the *Lisianthius skinneri* (Gentianaceae) species complex in Panama. They found restriction enzyme mapping to be a definitive help in resolving the phylogeny of this complex and that the rates of divergence differed markedly within and between lineages. They also found cpDNA mutation data to be congruent with rDNA phylogeny. However, rDNA analyses were found to have certain limitations. For example, it was not possible to distinguish between the two species *L. jefensis* and *L. skinneri*, which were clearly separable by isozyme analysis.

Doyle and Beachy (1985) have used restriction enzyme mapping to determine repeats and site locations for 18S and 25S rDNA genes in species of the genus *Glycine*. The authors found considerable differences between species which were congruent with those of previous biosystematic studies. Of interest was the finding that the rDNA of the cultivated soybean (*G. max*) was identical to its wild progenitor, *G. soja*, and no variation was found in over 40 accessions of the two species.

DeBonte *et al.* (1984) have carried out restriction analysis of both plastid and mitochondrial DNAs to aid in the classification of several species of *Daucus*. By this methodology plastid DNAs indicated a closer homology of *D. carota* subsp. *sativus* to *D. capillifolius* than to *D. carota* subsp. *gummifer* or *D. pusillus*. The mitochondrial restriction patterns also supported this close homology of *D. carota* subsp. *sativus* and *D. capillifolius*. In a somatic hybrid between *D. carota* and *D. capillifolius*, Matthews and Widholm (1985) reported that the plastid DNAs from the somatic hybrid and *D. carota* were identical, but that they were different from that of *D. capillifolius* when analysed on agarose electrophoretic gels after digestion by the restriction endonuclease HpaII. Other restriction enzymes did not show any differences between the two taxa. The results suggest that no major rearrangements of the plastid genome had occurred. In contrast, the mitochondrial DNA restriction patterns of *D. carota* and *D. capillifolius* were different from each other.

Hosaka *et al.* (1984) have classified the phylogenetic relationships of 37 species of *Solanum* and two species of *Lycopersicon* using eight restriction endonucleases for the analysis of chloroplast DNA. The authors found that *Solanum lycopersicoides* is different from other *Solanum* species by at least 21 mutational changes and is closely related to *Lycopersicon* species. They also found that the common potato is derived from Chilean *tuberosum*, whereas all other cultivated species have originated monophyletically from *Solanum stenotomum* (Fig. 7)

C. *DNA/DNA hybridization — nucleotide homologies*

Raina and Narayan (1984) have shown that the nuclear DNA amounts between three species of *Vicia* (*V. eriocarpa, V. johannis* and *V. melanops*) are the result of differences in the amounts of both repetitive and non-repetitive DNA sequences (Table 6). They consider that the increase in the non-repetitive fraction of DNA is due to a steady increase in highly diverged base sequences resulting from mutations, deletions, insertions and base sequence rearrangements among the repetitive sequences (Fig. 8). Some of the highly repetitive sequences have very low base sequence complexity, suggesting that these sequences have proliferated relatively recently during the evolution of the genus.

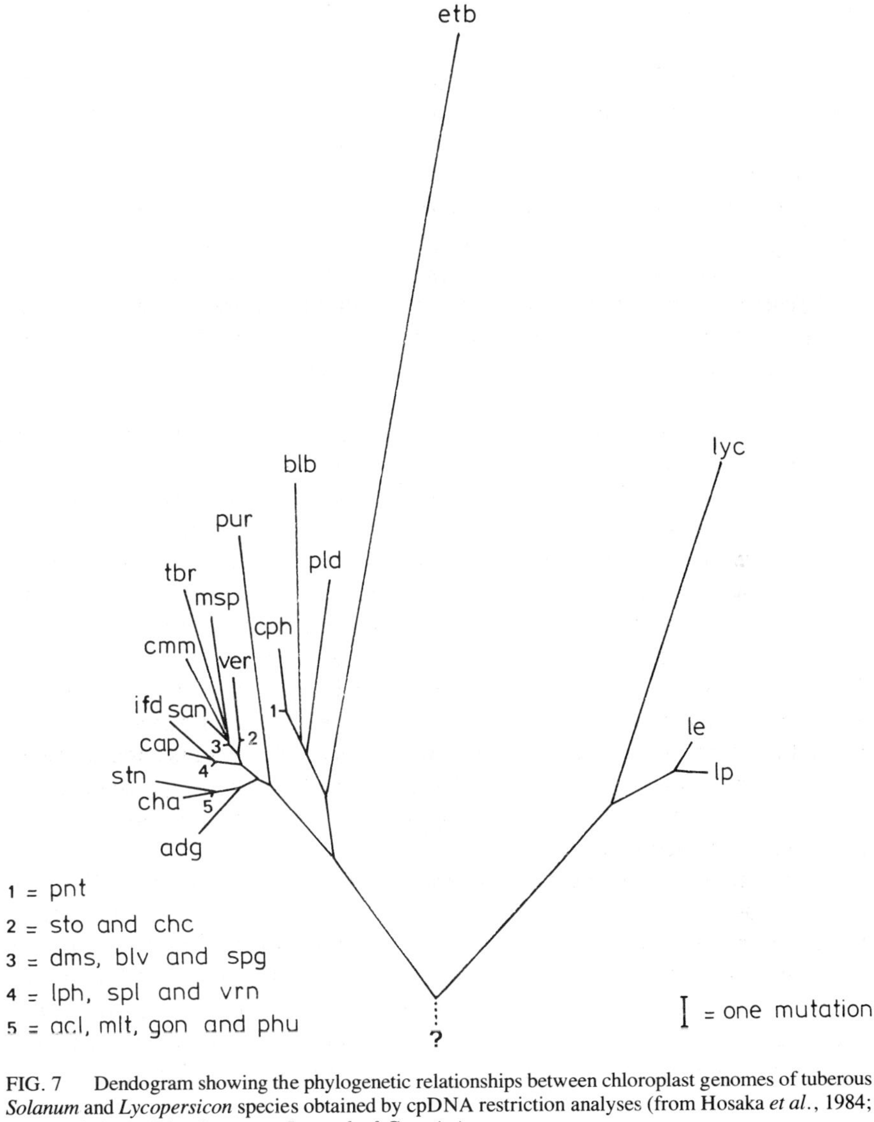

FIG. 7 Dendogram showing the phylogenetic relationships between chloroplast genomes of tuberous *Solanum* and *Lycopersicon* species obtained by cpDNA restriction analyses (from Hosaka *et al.*, 1984; by permission of the *Japanese Journal of Genetics*).

TABLE 6

Nuclear DNA composition in *Vicia* species in pg (modified from Raina and Narayan, 1984)

Species	Highly repetitive DNA		Mid-repetitive DNA		Non-repetitive DNA		Total DNA
	%	Amount	%	Amount	%	Amount	
V. eriocarpa	5	0.23	48	2.16	47	2.11	4.50
V. johannis	3	0.43	69	9.93	28	3.96	14.14
V. melanops	8	1.60	54	10.81	38	7.60	20.02

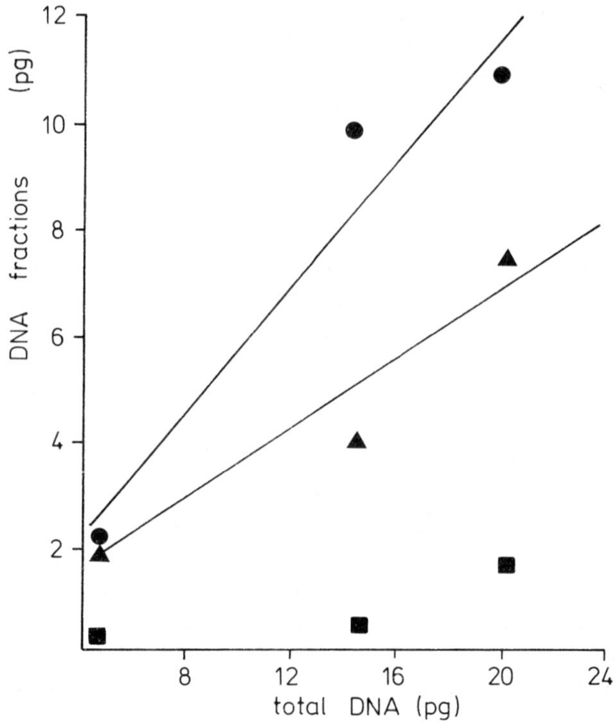

FIG. 8 The amounts of highly repetitive DNA (■), middle repetitive DNA (▲) and non-repetitive DNA (●) are plotted against the total DNA of the three *Vicia* species in Table 6 (from Raina and Narayan, 1984, with permission of Springer-Verlag and authors).

Schuler *et al.* (1983) have compared the nucleotide homologies and derived amino acid sequences of one of the genes encoding for seed storage protein of *Phaseolus vulgaris* and a conglycinin storage protein gene of *Glycine max*. From a comparison of the number of nucleotides inserted or deleted from the introns and the coding sequences, it was found that the nucleotide sequence of the genes of *G. max* and *P. vulgaris* overlap one another for a region of about 800 nucleotides corresponding to 28% of the *G. max* coding sequence and 38% of the *P. vulgaris* coding sequence. Both genes show extensive homologies which indicate that these genes have evolved from a common ancestral gene. *Phaseolus vulgaris* and *Glycine max* are members of different subtribes within the tribe Phaseoleae, and although rather closely related to one another, they still represent significant evolutionary divergence which is shown by the nucleotide comparisons.

While many additional examples could be given in which molecular techniques are now in use in aiding systematic research, one can conclude that the molecular field looks most promising for the future, and that such analyses will help resolve relationships that are not completely resolved by non-molecular approaches.

Although no one can doubt the potential of the molecular approach, we must heed the warning of John (1983) that it does not provide the whole story of development or evolution, as chromosomes regulate the recombination of genes by virtue of their behaviour at meiosis and fertilization.

V. Conclusions

The direct analysis of the genotype has been carried out on many levels — from the microscopic determination of chromosome numbers to more recent molecular analyses on the DNA level. Chromosome numbers still have a major role to play in resolving specific and generic affinities, polyploid levels and phylogenetic relationships. Our knowledge of chromosome numbers *per se* for the world's flora is still far from complete. For example, in the genus *Lotus* alone, it is estimated that it will require almost another century for the chromosome numbers to be determined for the remaining species of the genus. Likewise, nuclear DNA C-values are known for less than 1% of the angiosperm species. Nuclear DNA amounts have been considered constant for a species, but it is now known that wide intraspecific difference may occur. Total nuclear DNA content has been shown to have increased by equal increments of DNA to each chromosome independent of chromosome size. It may be that in the case of some high polyploids which have a lower DNA value than taxa on lower ploidy levels that the *decrease* is achieved by *losses* of equal increments of DNA from each chromosome. Many nucleotypic characters are now known to be associated with genome size (C-value) and are considered to be "independent" of coded informational content of the genome. The portrayal of the genome as a "natural karyotype" has led to a new understanding of karyotype architecture. Chromosome banding is a powerful tool for resolving karyotype and genome evolution. Molecular techniques, such as restriction mapping, for systematic and evolutionary studies, allow a direct analysis of the genotype rather than of derived phenotypes. A lateral interkingdom transfer of DNA from *Agrobacterium* to *Nicotiana* and its subsequent passage in evolutionary time in other species of *Nicotiana* has been reported. The understanding of such transformations will help resolve mechanisms of speciation and evolutionary pathways.

References

Abrahamson, S., Bender, M. A., Conger, A. D. and Wolff, S. (1973). Uniformity of radiation-induced mutation rates among different species. *Nature* **245**, 460–462.

Anderson, L. K., Stack, S. M., Fox, M. H. and Chuanshan, Z. (1985). The relationship between genome size and synaptonemal complex length in higher plants. *Exp. Cell Res.* **156**, 367–378.

Appels, R. and Dvorak, J. (1982). Relative rates of divergence of spacer and gene sequences within the rDNA region of species in the Triticeae: Implications for the maintenance of homogeneity of a repeated gene family. *Theor. Appl. Genet.* **63**, 361–365.

Armstrong, K. C. (1985). Chromosome pairing failure in an intersectional amphiploid of *Bromus altissimus* × *B. arvensis*. *Can. J. Genet. Cytol.* **27**, 705–709.

Bachmann, K., Chambers, K. L. and Price, H. J. (1985). Genome size and natural selection: Observations and experiments in plants. *In* "Evolution of Genome Size" (T. Cavalier-Smith, ed.), pp. 267–276. John Wiley, Chichester.

Baetcke, K. P., Sparrow, A. H., Naumann, C. H. and Schwemmer, S. S. (1967). The relationship of DNA content to nuclear and chromosome volumes and radiosensitivity (LD_{50}). *Proc. Natl Acad. Sci. U.S.A.* **58**, 533–540.

Bennett, M. D. (1971). The duration of meiosis. *Phil. Trans. Roy. Soc. Ser. B* **178**, 277–299.

Bennett, M. D. (1972). Nuclear DNA content and minimum generation time in herbaceous plants. *Proc. Roy. Soc. Ser. B* **181**, 109–135.

Bennett, M. D. (1976). DNA amount, latitude and crop plant distribution. *Environ. Exp. Bot.* **16**, 93–108.

Bennett, M. D. (1984). The genome, the natural karyotype, and biosystematics. *In* "Plant Biosystematics" (W. F. Grant, ed.), pp. 41–66. Academic Press, London and New York.

Bennett, M. D. (1985). Intraspecific variation in DNA amount and the nucleotypic dimension in plant genetics. *In* "Plant Genetics" (M. Freeling, ed.), pp. 283–302. Alan R. Liss, New York.

Bennett, M. D. and Smith, J. B. (1976). Nuclear DNA amounts in angiosperms. *Proc. Rpy. Soc. London Ser. B* **216**, 179–199.

Blakey, D. H. and Vosa, C. G. (1982). Heterochromatin and chromosome variation in cultivated species of *Tulipa* subg. *Keiostemones* (*Liliaceae*). *Plant Syst. Evol.* **139**, 163–178.

Bowman, C. M., Bonnard, G. and Dyer, T. A. (1983). Chloroplast DNA variation between species of *Triticum* and *Aegilops*. Location of the variation of the chloroplast genome and its relevance and classification of the cytoplasm. *Theor. Appl. Genet.* **65**, 247–262.

Brandham, P. E. and Bennett, M. D. (1983). "Kew Chromosome Conference II". George Allen & Unwin, London.

Brown, S. W. (1966). Heterochromatin. *Science* **151**, 417–425.

Butterfass, T. (1983). A nucleotypic control of chloroplast reproduction. *Protoplasma* **118**, 71–74.

Cavalier-Smith, T. (ed.) (1985a). "The Evolution of Genome Size". John Wiley, Chichester.

Cavalier-Smith, T. (1985b). Introduction: the evolutionary significance of genome size. *In* "The Evolution of Genome Size" (T. Cavalier-Smith, ed.), pp. 1–36. John Wiley, Chichester.

Cavalier-Smith, T. (1985c). Cell volume and the evolution of eukaryotic genome size. *In* "The Evolution of Genome Size" (T. Cavalier-Smith, ed.), pp. 105–184. John Wiley, Chichester.

Cavalier-Smith, T. (1985d). Eukaryote gene numbers, non-coding DNA and genome size. *In* "The Evolution of Genome Size" (T. Cavalier-Smith, ed.), pp. 69–103. John Wiley, Chichester.

Chandler, J. M., Jan, C.-C. and Beard, B. H. (1986). Chromosomal differentiation among the annual *Helianthus* species. *Syst. Bot.* **11**, 354–371.

Chinnappa, C. G. and Morton, J. K. (1978). Heterochromatin banding patterns in two species of *Trillium*. *Can. J. Genet. Cytol.* **20**, 475–481.

Clegg, M. T., Rawson, J. R. and Thomas, K. (1984). Chloroplast DNA variation in pearl millet and related species. *Genetics* **106**, 449–461.

Cortes, F., Gonzalez-Gil, G. and Hazen, M. J. (1983). C-banding and sister chromatid exchanges in three species of the genus *Allium* (*A. cepa, A. ascalonicum* and *A. sativum*). *Caryologia* **36**, 203–210.

Davies, D. R. (1977). DNA contents and cell number in relationship to seed size in the genus *Vicia*. *Heredity* **39**, 153–163.

DeBonte, L. R., Matthews, B. F. and Wilson, K. G. (1984). Variation of plastid mitochrondrial DNAs in the genus *Daucus*. *Am. J. Bot.* **71**, 932–940.

Doyle, J. J. and Beachy, R. N. (1985). Ribosomal gene variation in soybean (*Glycine*) and its relatives. Theor. Appl. Genet. **70**, 369–376.

Doyle, J. J., Beachy, R. N. and Lewis, W. H. (1984). Evolution of rDNA in *Claytonia* polypoloid complexes. *In* "Plant Biosystematics" (W. F. Grant, ed.), pp. 321–341. Academic Press, London and New York.

Doyle, J. J., Soltis, D. E. and Soltis, P. S. (1985). An intergeneric hybrid in the Saxifragaceae: Evidence from ribosomal RNA genes. *Am. J. Bot.* **72**, 1388–1391.

Dutta, S. K. (1986). "DNA Systematics", Vol. II: Plants. CRC Press, Boca Raton, Florida.

Edwards, G. A. and Endrizzi, J. E. (1975). Cell size, nuclear size and DNA content relationships in *Gossypium*. *Can. J. Genet. Cytol.* **17**, 181–186.

Ehrendorfer, F. (1983). Quantitative and qualitative differentiation of nuclear DNA in relation to plant systematics and evolution. *In* "Proteins and Nucleic Acids in Plant Systematics" (U. Jensen and D. E. Fairbrothers, eds), pp. 3–35. Springer, Berlin.

Evans, G. M. and Rees, H. (1971). Mitotic cycles in dicotyledons and monocotyledons. *Nature* **233**, 350–351.

Fivawo, N. C. and Rees, H. (1985). Chromosome size variation during pollen grain development in *Scilla sibirica*. *Theor. Appl. Genet.* **70**, 417–421.

Flavell, R. (1982). Sequence amplification, deletion and rearrangement: Major sources of variation during species divergence. *In* "Genome Evolution" (G. A. Dover and R. B. Flavell, eds), pp. 321–323. Academic Press, London and New York.

Flavell, R. (1983). Repeated sequences and genome architecture. *In* "Structure and Function of Plant Genomes" (O. Ciferri and L. Dure, eds), pp. 1–14. Plenum Press, New York and London.

Fukuda, I. (1984). Chromosome banding and biosystematics. *In* "Plant Biosystematics" (W. F. Grant, ed.), pp. 97–116. Academic Press, London and New York.

Fukuda, I. and Grant, W. F. (1980). Chromosome variation and evolution in *Trillium grandiflorum*. *Can. J. Genet. Cytol.* **22**, 81–91.

Fukui, K. (1986). Standardization of karyotyping plant chromosomes by a newly developed chromosome image analyzing system (CHIAS). *Theor. Appl. Genet.* **72**, 27–32.

Furner, I. J., Huffman, G. A., Amasino, R. M., Garfinkel, D. J., Gordon, M. P. and Nester, E. W. (1986). An *Agrobacterium* transformation in the evolution of the genus *Nicotiana*. *Nature* **319**, 422–427.

Gottschalk, W. (1985). Polyploidy and its role in the evolution of higher plants. *In* "Advances in Chromosome and Cell Genetics" (A. K. Sharma and A. Sharma, eds), pp. 1–43. Oxford, New Delhi.

Grant, W. F. (1969). Decreased DNA content of birch (*Betula*) chromosomes at high ploidy as determined by cytophotometry. *Chromosoma* **26**, 326–336.

Grant, W. F. (ed.) (1984). "Plant Biosystematics". Academic Press, London and New York.

Grant, W. F. (1986). The cytogenetics of *Lotus* (Leguminosae). *J. Nat. Hist.* **20**, 1461–1465.

Green, D. M., Myers, P. Z. and Reyna, D. L. (1984). CHROMPAC III: an improved package for microcomputer-assisted analysis of karyotypes. *J. Hered.* **75**, 143.

Grime, J. P. (1983). Prediction of weed and crop response to climate based upon measurements of nuclear DNA contents. *In* "Aspects of Applied Biology", Vol. 4: Influence of Environmental Factors on Herbicide Performance and Crop and Weed Biology, pp. 87–98. Association for Applied Biology, Nat. Veg. Research Station, Wellesbourne, Warwick.

Grime, J. P. and Mowforth, M. A. (1982). Variation in genome size, an ecological interpretation. *Nature* **299**, 151–153.

Gustafson, J. P. and Bennett, M. D. (1982). The effect of telomeric heterochromatin from *Secale cereale* on triticale (× *Triticosecale*). 1. The influence of the loss of several blocks of telomeric heterochromatin on early endosperm development and kernel characteristics at maturity. *Can. J. Genet. Cytol.* **24**, 83–92.

Hosaka, K., Ogihara, Y., Matsubayashi, M. and Tsunewaki, K. (1984). Phylogenetic relationship between the tuberous *Solanum* species as revealed by restriction endonuclease analysis of chloroplast DNA. *Jpn J. Genet.* **59**, 349–369.

Hsiao, C., Wang, R. R.-C. and Dewey, D. R. (1986). Karyotype analysis and genome relationships of 22 diploid species in the tribe Triticeae. *Can. J. Genet. Cytol.* **28**, 109–120.

Iltis, H. H. (1983). Tropical forests. What will be their fate? *Environment* **25**, 55–60.

Jansen, R. K. (1985). Systematic significance of chromosome numbers in *Acmella* (Asteraceae). *Am. J. Bot.* **72**, 1835–1841.

Jenkins, G. and Bennett, M. D. (1981). The intranuclear relationship between centromere volume and chromosome size in *Festuca scariosa* × *drymeja*. *J. Cell Sci.* **47**, 117–125.

John, B. (1983). Through the looking glass: a sceptical cytologist in a molecular wonderland. *In* "Kew Chromosome Conference II" (P. E. Brandham and M. D. Bennett, eds), pp. 305–310. George Allen & Unwin, London.

Karp, A., Rees, H. and Jewell, A. W. (1982). The effects of nucleotype and genotype upon pollen grain development in Hyacinth and *Scilla*. *Heredity* **48**, 251–261.

Kenton, A. (1984). Chromosome evolution in the *Gibasis linearis* group (Commelinaceae). III. DNA variation, chromosome evolution, and speciation in *G. venustula* and *G. heterophylla*. *Chromosoma* **90**, 303–310.

Lavania, U. C. and Sharma, A. K. (1985). Arrangement of the interphase chromosomes in the nucleus. *J. Hered.* **76**, 395–396.

Levin, D. A. and Funderberg, S. W. (1979). Genome size in angiosperms: temperate versus tropical species. *Am. Nat.* **114**, 784–795.

Linde-Laursen, I. (1984). Giemsa C-banding of barley chromosomes. IV. Chromosomal constitution of autotetraploid barley. *Hereditas* **100**, 219–223.

Linde-Laursen, I. and Jensen, J. (1984). Separate location of parental chromosomes in squashed metaphases of hybrids between *Hordeum vulgare* L. and four polyploid, alien species. *Hereditas* **100**, 67–73.

Linde-Laursen, I., von Bothmer, R. and Jacobsen, N. (1986). Giemsa C-banded karyotypes of *Hordeum* taxa from North America. *Can. J. Genet. Cytol.* **28**, 42–62.

Matthews, B. F. and Widholm, J. M. (1985). Organelle DNA compositions and isoenzyme expression in an interspecific somatic hybrid of *Daucus*. *Mol. Gen. Genet.* **198**, 371–376.

Martinez, A. and Ginzo, H. D. (1985). DNA content in *Tradescantia*. *Can. J. Genet. Cytol.* **27**, 766–775.

Murray, B. G. (1985). Karyotypes and nuclear DNA amounts in *Polypodium* L. (Polypodiaceae). *Bot. J. Linn. Soc.* **90**, 209–216.

Nagl, W. (1974). Role of heterochromatin in the control of the cell cycle duration. *Nature* **249**, 53–54.

Nagl, W. and Ehrendorfer, F. (1973). Chromosome size and DNA content in three species of Asteraceae-Anthemideae. *Oesterr. Bot. Z.* **121**, 165–169.

Nagl, W. and Ehrendorfer, F. (1974). DNA content, heterochromatin, mitotic index, and growth in perennial and annual Anthemideae (Asteraceae). *Plant Syst. Evol.* **123**, 35–54.

Newton, M. E. (1985). Heterochromatin diversity in two species of *Pellia* (Hepaticae) as revealed by C-, Q-, N- and Hoechst 33258-banding. *Chromosoma* **92**, 378–386.

Nurse, P. (1985). The genetic control of cell volume. *In* "The Evolution of Genome Size" (T. Cavalier-Smith, ed.), pp. 185–209. John Wiley, Chichester.

Okada, H. (1984). Polyphyletic allopolyploid origin of *Ranunculus cantoniensis* (4*x*) from *R. silerifolius* (2*x*) × *R. chinensis* (2*x*). *Plant Syst. Evol.* **148**, 89–102.

Palmer, J. D., Shields, C. R., Cohen, D. B. and Orton, T. J. (1983). Chloroplast DNA evolution and the origin of amphidiploid *Brassica* species. *Theor. Appl. Genet.* **65**, 181–189.

Perl-Treves, R., Zamir, D., Navot, N. and Galun, E. (1985). Phylogeny of *Cucumis* based on isozyme variability and its comparison with plastome phylogeny. *Theor. Appl. Genet.* **71**, 430–436.

Poggio, L. and Hunziker, J. H. (1986). Nuclear DNA content variation in *Bulnesia*. *J. Hered.* **77**, 43–48.

Price, H. J., Sparrow, A. H. and Nauman, A. F. (1973). Correlations between nuclear volume, cell volume, and DNA content in meristematic cells of herbaceous angiosperms. *Experientia* **29**, 1028–1029.

Price, H. J., Chambers, K. L. and Bachmann, K. (1981a). Genome size variation in diploid *Microseris bigelovii* (Asteraceae). *Bot. Gaz.* **142**, 156–159.

Price, H. J., Chambers, K. L. and Bachmann, K. (1981b). Geographic and ecological distribution of genomic DNA content variation in *Microseris douglasii* (Asteraceae). *Bot. Gaz.* **142**, 415–426.

Probst, F. (1972). Contribution à l'étude taxonomique du genre *Leucanthemum* par voie cytophotométrique. *Chromosoma* **36**, 322–328.

Raina, S. N. and Narayan, R. K. J. (1984). Changes in DNA composition in the evolution of *Vicia* species. *Theor. Appl. Genet.* **68**, 187–192.

Rayburn, A. L., Price, H. J., Smith, J. D. and Gold, J. R. (1985). C-band heterochromatin and DNA content in *Zea mays*. *Am. J. Bot.* **72**, 1610–1617.

Rees, H. (1984). Nuclear DNA variation and the homology of chromosomes. *In* "Plant Biosystematics" (W. F. Grant, ed.), pp. 87–96. Academic Press, London and New York.

Schlarbaum, S. E., Tsuchiya, T. and Johnson, L. C. (1984). The somatic chromosomes of *Cryptomeria japonica* with special reference to the marker chromosomes. *Silvae Genet.* **33**, 4–5.

Schuler, M. A., Doyle, J. J. and Beachy, R. N. (1983). Nucleotide homologies between the glycosylated seed storage proteins of *Glycine max* and *Phaseolus vulgaris*. *Plant Mol. Biol.* **2**, 119–127.

Sharma, A. K. and Sharma, A. (1985). "Advances in Chromosome and Cell Genetics". Oxford & IBH Publish., New Delhi.

Sims, L. E. and Price, H. J. (1985). Nuclear DNA content variation in *Helianthus* (Asteraceae). *Am. J. Bot.* **72**, 1213–1219.

Somaroo, B. H. and Grant, W. F. (1971). An interspecific diploid hybrid of *Lotus* (Leguminosae) with a B chromosome. *Can. J. Genet. Cytol.* **13**, 158–160.

Sorenson, J. C. (1984). The structure and expression of nuclear genes in higher plants. *Adv. Genet.* **22**, 109–144.

Stace, C. A. (1980). "Plant Taxonomy and Biosystematics". Edward Arnold, London.

Stebbins, G. L. (1976). Chromosome, DNA and plant evolution. *Evol. Biol.* **9**, 1–34.

Stern, D. B. and Palmer, J. D. (1984). Extensive and widespread homologies between mitochondrial DNA and chloroplast DNA in plants. *Proc. Natl Acad. Sci. U.S.A.* **81**, 1946–1950.

Sytsma, K. J. and Schaal, B. A. (1985). Phylogenetics of the *Lisianthius skinneri* (Gentianaceae) species complex in Panama utilizing DNA restriction fragment analysis. *Evolution* **39**, 594–608.

Taper, L. J. and Grant, W. F. (1973). The relationship between chromosome size and DNA content in birch (*Betula*) species. *Caryologia* **26**, 263–273.

Tateoka, T. (1985). Chromosome numbers of three species of *Agrostis* (Poaceae) in Japan. *Bull. Natl Sci. Mus., Tokyo, Ser. B* **11**, 33–40.

Trask, B., Van den Engh, G., Landegent, J., Jansen in de Wal, N. and Van der Ploeg, M. (1985). Detection of DNA sequences in nuclei in suspension by *in situ* hybridization and dual beam flow cytometry. *Science* **230**, 1401–1403.

Underbrink, A. G. and Pond, V. (1976). Cytological factors and their predictive role in comparative radiosensitivity: a general survey. *Curr. Topics Radiat. Res. W.* **11**, 251–306.

Van't Hof, J. (1965). Relationships between mitotic cycle duration, S period duration and the average rate of DNA synthesis in the root meristem of several plants. *Exp. Cell Res.* **39**, 48–58.

Van't Hof, J. (1975). The duration of chromosomal DNA synthesis, of the mitotic cycle, and of meiosis of higher plants. *In* "Handbook of Genetics 2" (R. C. King, ed.), pp. 363–377. Plenum Press, New York.

Vida, G. and Mohay, J. (1980). Cytophotometric DNA studies in polyploid series of the fern genus *Cystopteris* Bernh. *Acta Bot. Acad. Sci. Hung.* **26**, 455–461.

Vosa, C. G. (1985). Chromosome banding in plants. *In* "Advances in Chromosome and Cell Genetics" (A. K. Sharma and A. Sharma, eds), pp. 79–104. Oxford and IBH Publishing, New Delhi, Bombay, Calcutta.

Werth, C. R., Guttman, S. I. and Eshbaugh, W. H. (1985). Recurring origins of allopolyploid species in *Asplenium*. *Science* **228**, 731–733.

Wilson, E. O. (1985). Biophilia. Time to revive systematics. *Science* **230**, 1227.

Zinov'eva-Stahevitch, A. E. and Grant, W. F. (1984a). Chromosome numbers in *Impatiens* (Balsaminaceae). *Can. J. Bot.* **62**, 2630–2635.

Zinov'eva-Stahevitch, A. E. and Grant, W. F. (1984b). Polysomaty in the genus *Impatiens*. *Caryologia* **37**, 435–438.

Zinov'eva-Stahevitch, A. E. and Grant, W. F. (1985). Chromosomal differentiation in *Impatiens* L. (Balsaminaceae). *Caryologia* **38**, 139–145.

2 Pollen Differentiation Patterns and Male Function

R. BRUCE KNOX

Plant Cell Biology Research Centre, School of Botany, University of Melbourne, Parkville, Victoria, Australia

I. Introduction

A major problem in plant reproductive biology today is that the seed production in plants may be greatly reduced by failure to achieve full reproductive capacity. Reduced seed output may occur in different species in the field, or in greenhouse or growth chambers, in both natural and experimental populations of plants. In order to control the processes of fertilization, a great deal of information is needed concerning the interacting partners, especially in terms of the male programme that produces the viable pollen grain as well as the molecular nature of communication between pollen and the receptive pistil.

A useful perspective on male reproductive function can be obtained from the viewpoints of population biology, in terms of two recently developed concepts: (1) *reproductive effort*, in which the plant's resources devoted to this process are estimated; and (2) *reproductive success* of the plant in parenting fertilized ovules (see review by Willson, 1983). These two concepts provide a theoretical basis for focusing our laboratory technologies.

In our laboratory we have set up a concentration of technologies that can be applied to plant reproductive biology: light and electron microscopy; image analysis, including three-dimensional reconstruction with interactive computer graphics; cytochemistry; embryo rescue and cloning; thermogravimetry; pollination and breeding technology; pollen freezing and storage; cell fractionation; monoclonal antibody technology; mentor and pioneer pollen technology; and

Differentiation Patterns in Higher Plants
ISBN: 0-12-709425-3

approaches to gene technology. These methods provide essential background information and data from which it is possible to design strategies both to solve reproductive problems in crop plants, and to explore the reproductive biology of wild populations of plants.

II. Male Reproductive Effort

The concept of male reproductive effort is based on the amount of the plant's total resource budget devoted to the fruit and seed set programme. It is subject to environmental factors and influenced by natural selection (see review by Willson, 1983). Reproductive effort is a key parameter, since the resources available may change diurnally, seasonally and annually, and also at any period of pre- or post-fertilization development. Fertilization is seen as only one of the events in seed-setting. Reproductive effort is a factor maximizing fitness or fecundity, i.e. the rate at which an individual plant contributes genes to future generations.

Reproductive effort can be estimated by the size and mass of seeds produced by a given pollen parent, or in physiological studies by the estimation of the metabolic cost of reproduction. The whole process of reproduction can be terminated at any point, e.g. if a particular stress is applied to the plant, the fruit or seed-set programme can be arrested, and seeds aborted or fruits shed precociously.

III. Male Reproductive Success

The success of an individual plant is estimated in terms of fathering as many viable embryos as possible (see Willson, 1983). The concept includes:

- pollen quantity;
- pollen viability or quality;
- effectiveness of the male gametes;
- pollen gene expression;
- the male programme of ontogeny.

This parameter differs from reproductive effort in that the influence of each individual component can be precisely defined experimentally in relation to controls so that the effects of reproductive effort can be excluded.

A. Pollen quantity

This useful quantitative character is seldom computed by reproductive biologists, far less by systematists. There are two interesting aspects that can be illustrated from studies of the Australian species of *Acacia* which have compound pollen or polyads. First, the pollen output can vary strikingly between closely related species. For instance, *Acacia myrtifolia* has three-flowered heads each bearing 12 900 polyads, while *A. retinodes* has 25-flowered heads, each bearing 13 200 polyads. The difference here is in the number of anthers per flower, which is about 530 for the former and 66 for the latter species. Second, the number of grains per polyad is correlated with the number of ovules in the same species. There is a polymorphism for grain number, varying from 4 to 16 in Australian species. The ovule number is precisely related but usually less than the polyad grain number, e.g. 1–3 for 4-ad species, 5–9 for 8-ad species, 7–9 for 12-ad species, and 9–14 for 16-ad species (Kenrick and Knox, 1982).

This factor has been calculated in a few plants in connection with the pollen–ovule ratio, but is rarely cited directly (e.g. Cruden, 1977). Pollen quantity has also been calculated in different grass species in establishing their allergenic potential (Smart *et al.*, 1979). *Lolium perenne* is among the highest, with large numbers of pollen grains per anther and per flowering spike. Pollen output on a per hectare basis approaches half a tonne in Melbourne if the plants are not mown or grazed (Smart *et al.*, 1979); the male reproductive success of this species is obviously very high. As with other exotic pasture grasses, pollen quantity is higher than that in endemic Australian grasses, e.g. *Themeda*. Pollen quantity may provide a useful quantitative character that has to date been largely ignored.

B. Pollen viability and quality

Two important aspects need to be considered:

- pollen viability, defined as ability of pollen to carry out successful fertilization and seed setting (Heslop-Harrison *et al.*, 1984);
- pollen quality, defined as having the properties of living cells, e.g. being osmotically active, or possessing an effective plasma membrane.

The distinction that is important, in terms of male function, is obviously ability to fertilize. This can be tested quite simply by seed-set experiments; however, they are subject to hazards such as self-incompatibility, stigma receptivity, ovule viability, etc., that can render interpretation equivocal. Heslop-Harrison *et al.* (1984), who point out these difficulties, recommend use of the FCR test based upon the pollen membrane integrity. Where the FCR test can not be used, there are many other tests, including pollen germination tests and a non-destructive test involving use of NMR spectroscopy (see review by Knox, 1984a,b). Unfortunately, these tests give only a statistical result, indicating the percentage of living grains. The question needs to be asked whether the dead grains present are the result of the operation of environmental factors or pollen sterility genes? Most of these genes, on the one hand, act at the microsporocyte stage before meiosis, but some act on the young microspores. In some systems, the precise period when the gene or its product acted and growth ceased, may be determined by estimating the microspore wall diameter. Breakdown and cell death can be assumed to have occurred at the developmental period corresponding to the final microspore dimensions. Environmental factors, on the other hand, may act at any stage of differentiation, e.g. through low or high temperature effects. A consequence is that even mature pollen, perfectly formed, can be rendered inviable. Recently, experiments on pollen death have shown that an early response is the loss of cellular membranes (Beattie *et al.*, 1984).

C. The effectiveness of the male gametes

Reproductive success is critically dependent on the effectiveness of the pair of male gametes in each pollen grain. During the past five years, knowledge of their structure and function has been transformed with two important discoveries:

- the physical association in mature tricellular pollen of the pair of sperm cells and vegetative nucleus, the *male germ unit* (Dumas *et al.*, 1984a,b);
- dimorphism of the pair of sperm cells involving an unequal distribution of heritable cytoplasmic organelles and differences in size of the sperm cells (Russell and Cass, 1983; Russell, 1984).

The *male germ unit* concept was developed to explain the physical association of all the DNA of male heredity. The pair of sperm cells may be so placed or possess cell surface determinants to participate in preferential fertilization, that is, the sperm cells may exhibit specificity in fusing with one or other of the female target cells — the egg or central cell. Detailed ultrastructural studies in *Plumbago zeylanica* have demonstrated cytoplasmic heterospermy (Russell, 1985). The sperm cell not attached to the vegetative nucleus, S_{ua}, is mitochondrion-poor but plastid-rich, and shows a high probability of fusing with the egg cell to form the embryo. In contrast, the other sperm cell that is attached to the vegetative nucleus, S_{vn}, is mitochondrion-rich but plastid-poor, and shows a high probability of fusing with the central cell to form the endosperm. The male germ unit has been demonstrated unequivocally by three-dimensional reconstruction technology in several dicotyledons: *Plumbago Zeylanica* (Russell, 1984), *Brassica campestris* (McConchie *et al.*, 1985; see Fig. 1), *B. oleracea* (Dumas *et al.*, 1985; McConchie *et al.*, submitted) and *Spinacia oleracea* (Wilms and van Aelst, 1983). Observations of two-dimensional projections indicate that such associations may also occur in *Gossypium hirsutum* (Jensen and Fisher, 1968). Among monocotyledons the sperm cells appear physically separate, although sited close to the vegetative nucleus in mature pollen of barley, *Hordeum vulgare* (Mogensen and Rusche, 1985) and *Zea mays* (McConchie *et al.*, 1987). However, once the generative cell enters the pollen tube, the situation can change; physical associations have been reported in several dicotyledon systems including *Petunia* (Wagner and Mogensen, pers. comm.) and *Rhododendron* (Kaul *et al.*, in press), and in one monocotyledon system *Hippeastrum* (Mogensen, 1986).

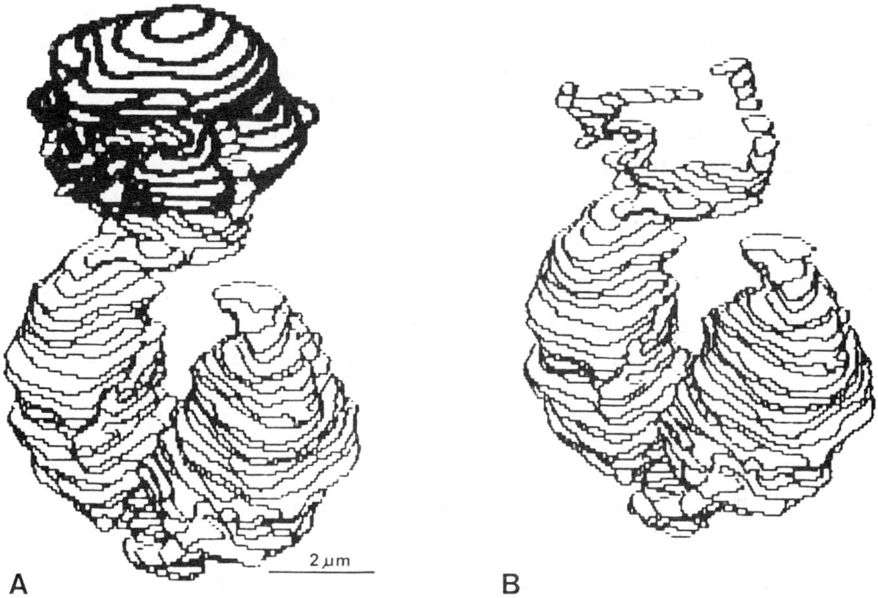

FIG. 1 *Brassica campestris*: three-dimensional reconstruction of the male germ unit in pollen. (A) The entire unit, comprising the vegetative nucleus (top of picture, bold print) and the pair of sperm cells: S_{ua} on right, attached only to its partner; S_{vn} on left, which has an extension whose branches penetrate into enclaves in the vegetative nucleus. (B) The same unit, but without the data for the vegetative nucleus so that the branched extensions of S_{vn} are clearly visible (top of picture). These wire-frame diagrams are produced using the Zeiss-Kontron Image Analysis System, and each contour represents the outline of a thin section, with the hidden lines removed. From McConchie *et al.* (1985).

In *Brassica campestris*, three-dimensional reconstruction technology is being applied to follow the ontogeny of the generative cell and its division to form the sperm cells for two purposes:

- to detect when the physical association is first initiated during ontogeny;
- to monitor the changes in number of heritable cytoplasmic organelles.

Three-dimensional reconstruction technology is based on the use of interactive computer graphics to generate images of the surfaces of the cells. If any ontogenetic sequence is desired, a series of models based on reconstruction of serial thin sections can be prepared. These will illustrate some of the ontogenetic changes, and the potential exists to generate artificial images corresponding to intermediates that may predict the situation in life.

The relative numbers of heritable organelles can be included in the reconstructions, providing topographical and numerical information. *B. campestris* is known to have the lowest number of mitochondria reported for the S_{ua} in angiosperms, namely four, while *B. oleracea* has six (McConchie *et al.*, 1987). The question arises whether such small numbers reflect the partitioning of cytoplasm at generative cell division or a progressive reduction during sperm cell differentiation, e.g. by organellar diminution as has been reported in barley (Mogensen and Rusche, 1985). The ontogeny of the sperm cells will, in the future, provide interesting characters of importance in understanding the nature of double fertilization in angiosperms.

Further insight will be obtained from studies of the cell biology of the sperm cells, especially the nature of any cell surface determinants. One way of obtaining such information is by using cell fractionation technology for sperm cell isolation. The isolated cells can then be characterized by a variety of biochemical procedures, including monoclonal antibodies. The feasibility of this approach has recently been demonstrated by the isolation of sperm cells of *Brassica* and *Zea* (Cass *et al.*, 1986) and *Plumbago* (Russell, 1986). Hybridoma cell lines have been produced against sperm cells of *Brassica* (Hough *et al.*, 1986). The goal of such studies is to determine whether sperm:egg recognition is mediated by specific surface membrane determinants, and ultimately to establish the mechanism of double fertilization.

IV. Pollen Gene Expression

What is known of the genes that are expressed during pollen development? This question has only recently been approached. In studies of maize and *Tradescantia* pollen, Mascarenhas *et al.* (1984) have shown that about 75% of the genes of the sporophyte are expressed in the pollen. Is there evidence of post-meiotic transcription of the haploid genome? At present, only the following three cases of biochemically detectable mutations segregating in pollen are known:

- *waxy* and normal starch in maize (Brink and MacGillivray, 1924; Demerec, 1924);
- presence or absence of alcohol dehydrogenase (*Adh*) in maize (Freeling, 1976);
- presence or absence of β-galactosidase (*gal*) in rapeseed (Singh and Knox, 1984, 1985).

The two enzymic markers are first expressed in the young microspores (Stinson and Mascarenhas, 1985; Singh *et al.*, 1985). Activity of both enzymes is absent in the microsporocytes, tetrads of microspores, and at spore release, but begins to build up in the young microspores (Fig. 2). In the case of *Adh*, there is only a single peak of enzymic activity, while in β-galactosidase, there is evidence for two peaks.

These biochemical markers provide temporal characters of microspore differentiation, as mutant pollen that is defective in enzymic expression, is apparently functional. In the case of β-galactosidase, male transmission of the mutant gene *gal* is reduced, so that pollen carrying this gene is less effective in fertilization (Singh and Knox, 1985).

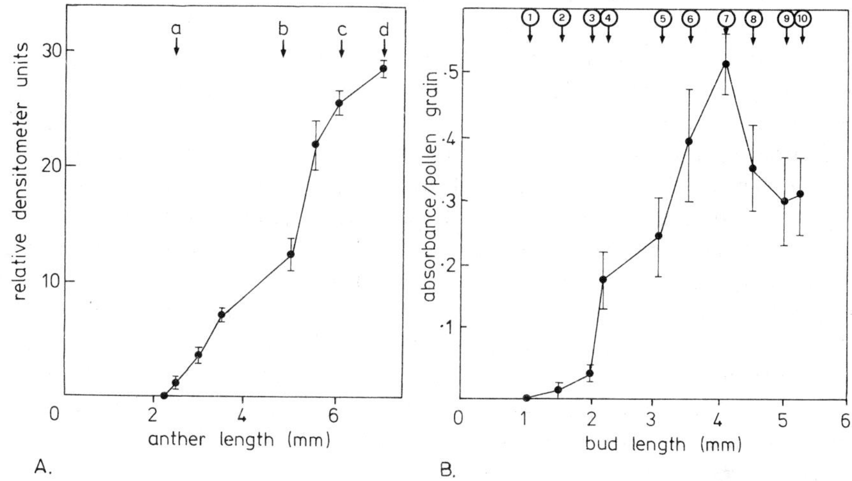

FIG. 2 (A) Changes in ADH activity during microsporogenesis of maize pollen; average values of three to five grains ± SE; a, tetrads; b, microspore mitosis; c, generative cell division; d, mature pollen (Stinson and Mascarenhas, 1985). (B) Quantitative microspectrophotometry of β-galactosidase in developing microspores and pollen of *B. campestris*. Mean absorbance per pollen grain (± 1 SE) is plotted against bud length (Singh *et al.*, 1985).

V. The Male Programme of Ontogeny

Any serious study of pollen ontogeny needs to incorporate certain features in its experimental design:

- identification of key periods in developmental sequences;
- definition of morphological or morphometric criteria to define each period;
- reproducibility of sampling procedures in different genotypes or seasons;
- estimation of pollen viability before sampling at each period.

While early studies in this field succeeded in unravelling the principal cytological features of pollen development, they rarely took the above criteria into account. They provide a series of sketches based on interpretation of semi-thin sections, showing cytological diagrams placed together into developmental sequences on the basis of pollen diameter or anther size (for review, see Maheshwari, 1950).

In the 1960s, ultrastructural studies continued this tradition, with emphasis being placed on the need to comprehend the complexities of wall differentiation (see Heslop-Harrison, 1968). Extensive studies by several groups centred around wall differentiations in *Lilium* pollen. Following the biochemical studies of Stern (1966), a relationship between anther length and microspore development was demonstrated using cytological staining techniques.

Later, a new technology was applied to the problem of estimating microspore and pollen viability with the application of the FCR (fluorochromatic reaction test,

Heslop-Harrison and Heslop-Harrison, 1970). For the first time live pollen grains could be directly tested within a few minutes, and the result scored by fluorescence microscopy. A goal was achieved when the technique was adapted for identifying developmental periods of *Lilium* and other pollen types (Knox and Heslop-Harrison, 1970). The method is based on a combination of microspore or pollen morphology as seen by phase contrast optics, and appearance of the large central vacuole by FCR and fluorescence microscopy. The vacuole is characteristic of the major period of microspore growth prior to pollen mitosis, known in classical cytology as the "signet-ring" period.

These studies enabled the following sequential events of pollen ontogeny to be determined:

- appearance of tetrad of microspores;
- release of young microspores from the callose special wall;
- vacuolation and growth of the microspores;
- asymmetric microspore mitosis cutting off the small generative cell from the large vegetative cell;
- pollen maturation, and in some systems sperm cell formation, and the association dehydration of the grain.

The same techniques permitted, for the first time, determination of the duration of each of the defined stages over a thirteen day period in *Phalaris aquatica* (Vithanage and Knox, 1980). Recently, a second fluorescent probe, the specific DNA fluorochrome DAPI, has been combined with the FCR test to detect pollen development periods in *Brassica campestris* (Singh *et al.*, 1985). With this method, the nuclear division patterns may be established in whole mounts of microspores or pollen. In *Brassica*, anthers or microspores were chemically fixed before DAPI fluorescence determinations. However, Vergne *et al.* (in press) have described the use of DAPI in a permeabilization medium which enables the fluorochrome to be used with live pollen. Such methods are rapid, taking only a few minutes to perform. A disadvantage is that the method requires fluorescence microscopy, and thus is not a technique readily applicable for field use.

Morphometric methods have also been developed. Perhaps the most elegant example is that for *Lilium* based on estimates of cell and wall volume (Willemse and Reznickova, 1980). The same authors have detailed the ultrastructural characteristics of each period, and have extended their studies to other systems, including *Gasteria* (M.T.M. Willemse, pers. comm., 1985). The terminology used, however, is naturally complex. Willemse and Reznickova (1980) identified six periods of microspore development, and eight periods of pollen differentiation. An unfortunate effect of their reliance on volume changes is that some of the periods are characterized essentially on a time basis, e.g. 4, 2 and 1 days before dehiscence.

To summarize these new technologies and their achievements, we can conclude:

- organization of the tetrad of microspores has been advanced by use of the aniline blue fluorescence method;
- cytoplasmic vacuolar changes have been visualized by use of the FCR test;
- the patterns of nuclear division have been determined by use of DNA fluorochromes.

The question needs to be asked whether there is any simpler way of expressing our knowledge of pollen development that may be useful, not only in studies of cell biology, but may also provide much needed data to develop character analysis of pollen phylogenies? The potential of the use of phylogenetic tools of analysis have begun to be explored for wall characters during microspore and pollen ontogeny (Blackmore and Crane, in press). There appears to be enormous potential for the use of cladistics in analysing ontogenetic characters, following the examples for animal systems by Wiley (1981).

A. Characters of pollen ontogeny

The concept of ontogenetic precedence was first outlined by Hennig (1966) for animal systems. Later, Nelson (1978) developed a notation to designate changes to homologous characters. Applying this system to microspore and pollen ontogeny, five characters of microspore ontogeny and six of pollen can be defined in two monocotyledon and two dicotyledon systems. Each follows sequentially, suggesting it is epigenetic, and each character with a particular number appears to be directly comparable to that in other taxa.

An objective of evolutionary studies of ontogeny is to search out shared advanced or derived characters (synapomorphies). These can be characteristic of and define groups, as well as specifying the probable pathway of evolution (Nelson, 1978). It follows, therefore, since the general character transforms into the less general, that the general character is ancestral, and the less general more advanced and specific to that group. Can we apply this concept to the characters of pollen ontogeny?

1. Microspore character analysis. The ontogenies of two monocotyledons, the grasses *Lolium perenne* and *Phalaris aquatica* are well established (Vithanage and Knox, 1980; Knox, 1984a). The microspores (Fig. 3a) show the progression established initially for *Lilium* (Knox and Heslop-Harrison, 1970) from tetrad of microspores, free microspores, vacuolation and growth and finally microspore mitosis.

$$\text{LP/PA (M)} \quad 1 \rightarrow 2 \rightarrow 3 \rightarrow 4$$

The microspore characters are referred to by the following notation:

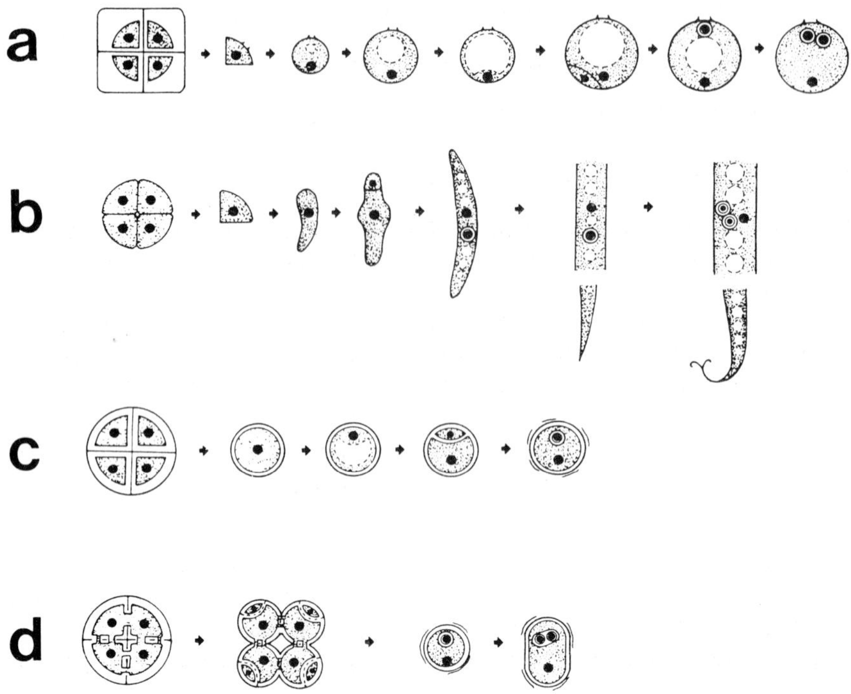

FIG. 3 Ontogeny of pollen of various monocotyledons. (a) *Lolium perenne* and *Phalaris aquatica*; (b) *Amphibolis antarctica*; (c) *Lepilaena cylindrocarpa*; (d) *Najas tenuifolia*. Interpretative sketches based on data of (a) Vithanage and Knox (1980); (b) Ducker *et al.* (1978); (c) McConchie and Knox (submitted a); (d) McConchie and Knox (submitted b).

1 *tetrad of microspores* held by intra-tetrad connections within the callose special wall;
2 *microspore* release as free non-vacuolate cells;
3 *microspore vacuolation* during major growth period;
4 *microspore mitosis*.

The ontogeny of microspores of two related species of dicotyledons, namely rapeseed, *Brassica campestris* and kale, *B. oleracea* has also been established (Singh *et al.*, 1985).

This may be represented in terms of the above microspore characters as follows:

$$\text{BC(M)} \quad 1 \rightarrow 2 \rightarrow 3 \rightarrow 4$$

A more complex situation apparently exists in systems with compound pollen, e.g. *Acacia retinoides* which has its microspores arranged in flattened disks of 16 grains. These arise within a single loculus of the anther by the cohesion of the products of four sporogenous cells (Newman, 1933; Kenrick and Knox, 1979; Knox and Kenrick, 1983) and their ontogeny may be represented as follows:

AR(M) $1' \rightarrow 2'' \rightarrow 3 \rightarrow 4$

The non-terminal characters are referred to as follows:

1′ tetrads of microspores held by inter-tetrad connections within the callose special wall;
2″ microspores not released from tetrad.

The notation used includes a single prime symbol (′) to indicate modifications or alternations to characters, and a double prime (″) to indicate negative effects or absence of the character.

During ontogeny of the compound microspores, cohesion is achieved in the normal way within each tetrad (character 1), but the compound state is assured by inter-tetrad connections between the four tetrads (character 1′). Later in ontogeny, these cytoplasmic connections are replaced by wall links that prevent release of the microspores (character 2″). The ultrastructural features of these events have been described by Knox and McConchie (1986).

2. *Pollen character analysis.* Following the microspore mitosis, the pollen grains are bicellular and comprise a generative cell and a vegetative cell. This is essentially the condition in which 70% of pollen species are shed. Surprisingly enough, several periods are readily identified in pollen of grasses *Lolium perenne* and *Phalaris aquatica*:

LP/PA(P) $5 \rightarrow 6 \rightarrow 7 \rightarrow 8$ (see Fig. 3a)

The pollen characters are referred to as follows:

5 generative cell attached to pollen wall;
6 generative cell rounds off, within vegetative cell cytoplasm;
7 generative cell division forms pair of sperm cells;
8 vegetative cell accumulates reserves;
9 germinal aperture present.

In the pollen of *Brassica* the sequence differs in the terminal characters:

BC(P) $5 \rightarrow 6 \rightarrow 7' \rightarrow 8'$

This sequence shows two additional terminal character changes, namely:

7′ sperm cells form male sperm unit together with vegetative nucleus; and
8′ vegetative cells accumulate reserves but starch is hydrolysed to give lipid-rich grains.

The notation can also be adapted for other terminal characters, e.g. in *Acacia retinoides* polyads, unusual membrane labyrinths differentiate at the apertures (McCoy and Knox, submitted). These are period specific, first developing at female phase of flowering. Only 24 hours later, at male phase, the labyrinths are filled with coated vesicles, apparently a mechanism for protein transport at the germinal apertures. A curious feature of *Acacia* pollen is that the generative cell apparently remains attached to the vegetative cell plasma membrane. The notation for *Acacia* is:

$$\text{AR(P)} \quad 5 \rightarrow 6'' \rightarrow 8 \rightarrow 9'$$

where an additional terminal character, 9′, represents the unusual apertural labyrinths in contrast to the normal apertures (9). The non-terminal additional character, 6″, indicates that the generative cell does not round off, but remains attached to the wall of the vegetative cell (McCoy and Knox, submitted). The question now arises as to whether this approach has applications in studying pollen phylogenies in a natural group of taxa. Unfortunately, few studies have been carried out on such a basis, but a number of ontogenetic studies have recently been completed of pollen of the aquatic monocotyledon group, the Helobiae (Alismatidae), which show the potential of this approach.

B. Case studies: the characters of pollen ontogeny in the Helobiae (Alismatidae)

This group of mainly aquatic monocotyledons comprises eight families with more than 26 genera (Tomlinson; in Metcalfe, 1982). Detailed studies of pollen ontogeny in three Australian species, using light and electron microscopy, have given sequences of characters with marked variation from the theme established above for terrestrial types.

The Australian sea nymph, *Amphibolis antarctica*, a dioecious seagrass of the family Cymodoceaceae, is unusual in several characters:

- pollen is filiform, each grain is up to 5.000 μm in length (Ducker and Knox, 1976);
- the outer wall layer, the exine is completely absent, as judged by light microscope cytochemistry (Ducker and Knox, 1976) and transmission electron microscopy (Ducker *et al.*, 1978);
- mature pollen is vacuolate.

During ontogeny, there are three other important characters (Fig. 3b):

- the tetrad of microspores is not surrounded by the classical callose special wall, but held internally by a crucifix of callose between each microspore (Ducker *et al.*, 1978);
- immediately following the tetrad period, the free microspores undergo precocious mitosis to form the generative cell prior to the period of vacuolation and growth (Ducker *et al.*, 1978);
- the free microspores are initially bounded only by their plasma membrane, the wall being initiated later at vacuolate period when intine synthesis usually begins in other systems (Pettitt *et al.*, 1984).

The sequence of microspore characters is therefore the following:

AA(M) 1″→ 2→ 3″→ 4

Character 1″ indicates that the microspores are not enclosed within a callose special wall. Character 3″ indicates that the microspore fails to undergo vacuolation. Pollen differentiation thus occupies the greater part of anther development:

AA(P) 5→ 3′→ 6→ 7→ 8→ 9″

A non-terminal character change has occurred, represented as follows:

3′ vegetative cell undergoes vacuolation.

The terminal character 9″ indicates that the pollen has no pre-formed apertures. Character 3′ is obviously analogous to character 3, merely delayed in its expression. The change in shape to filiform is unusual, and occurs almost immediately following microspore division. The filiform shape is characteristic of certain genera of Helobiae and in certain tropical families, e.g. Acanthaceae (Brummitt *et al.*, 1980).

The mature pollen of *Lepilaena cylindrocarpa* of the family Zannichelliaceae, is spherical, 40–45 μm in diameter, and covered by a reticulate net-like exine (McConchie *et al.*, 1982; Vierssen *et al.*, 1982). Its ontogeny is shown in Fig. 3c (McConchie, 1983; McConchie and Knox, submitted a), and is unusual among Helobiae in being bicellular at maturity. The sequence of characters is otherwise typical of those outlined above for grasses, and is as follows:

LC(M) 1→ 2→ 3→ 4 LC(P) 5→ 6→ 7″→ 8→ 9″

The character 7″ indicates that sperm cell division does not occur in the pollen grain.

The third case is *Najas tenuifolia* of the family Najadaceae. The mature pollen is elliptical in shape, this feature apparently being determined relatively late in development following dissolution of the periplasmodial tapetum, and prior to sperm cell division (McConchie, 1983; McConchie and Knox, submitted b). Its ontogeny is unusual in other respects (Fig. 3d):

- tetrads of microspores persist, linked by cytomictic channels, until mid-bicellular pollen period when the tetrads separate into monads; precocious microspore mitosis occurs within the tetrad period;
- absence of a vacuolate period.

The ontogenetic sequence of characters is therefore as follows:

NT(M) 1→ 2″→ 3″→ 4

The non-terminal microspore character change occurred here as follows:

2″ microspores remain coherent in tetrads (as in *Acacia* above).

The notation for pollen development may be expected to be unusual in the light of these ontogenetic characters, and is as follows:

$$\text{NT(P)} \quad 5\rightarrow 3'\rightarrow 6\rightarrow 2'\rightarrow 7\rightarrow 8\rightarrow 9''$$

The new terminal and non-terminal characters indicate:

2′ pollen tetrads separate into monads;
3′ vegetative cell remains non-vacuolate.

To summarize, pollen ontogenies can readily be given a simple notation that represents character sequences. For the monocotyledons, these are as follows (using symbols as described above):

PA/LP(M)	$1\rightarrow 2\rightarrow 3\rightarrow 4$	(P)	$5\rightarrow 6\rightarrow 7\rightarrow 8\rightarrow 9$
LC(M)	$1\rightarrow 2\rightarrow 3\rightarrow 4$	(P)	$5\rightarrow 6\rightarrow 7''\rightarrow 8\rightarrow 9''$
AA(M)	$1''\rightarrow 2\rightarrow 3''\rightarrow 4$	(P)	$5\rightarrow 3'\rightarrow 6\rightarrow 7\rightarrow 8\rightarrow 9''$
NT(M)	$1\rightarrow 2''\rightarrow 3''\rightarrow 4$	(P)	$5\rightarrow 3'\rightarrow 6\rightarrow 2'\rightarrow 7\rightarrow 8\rightarrow 9''$

This notation works well, and changes are made evident by the prime symbol. It is possible to indicate the timing of the ontogenetic sequence in the case of *Phalaris* (PA), where the data are available (Vithanage and Knox, 1980).

The examples presented above demonstrate that variation in ontogenetic characters of pollen development does occur in closely related families. Changes of adaptive significance have arisen, perhaps associated with aquatic pollination syndromes. From an epigenetic viewpoint, it is evident that, as with variation in form and function of angiosperm stigmas (Knox, 1984b), nature has adopted almost every conceivable variation in the basic strategy plan of pollen ontogeny. More effort now needs to be made to determine the usefulness of these epigenetic characters in understanding pollen phylogenies. The characters described above are important cytological or morphological differences in ontogenetic sequences. Most are causal, in the sense of Alberch (1985), in that they are required for normal pollen viability. An exception may be vacuolation of the cytoplasm — which is not required in *Najas*, as shown above.

Since in *Amphibolis* the vacuolate period is characteristic of pollen rather than microspore development, this character may be temporal rather than causal.

VI. Conclusions

We can conclude from this review that both the necessary technologies and theoretical basis presently exist for character analysis of pollen phylogenies. What is needed are more experimental data.

Technologies needed to obtain such data include microscopic analysis:

- Nomarski DIC optics for study of cytological changes during microspore and pollen ontogeny of whole microspores or grains, if necessary after appropriate clearing to reveal internal structure;

TABLE 1

List of characters of the male programme, as defined by the microscopic techniques outlined in the text. The list is not intended to be exclusive, and additional characters or character states can be added

A. Microspore programme

1 Tetrad of microspores cohere by means of intra-tetrad connections within the callose special cell wall;
OR
1′ Tetrads cohere by inter-tetrad connections within special cell wall to form polyads;
OR
1″ Tetrad not enclosed by callose special wall.

2 Microspore release giving free non-vacuolate microspores;
OR
2″ Microspores not released from tetrads.

3 Microspore vacuolation during major growth period;
OR
3″ Microspore fails to vacuolate.

4 Microspore mitosis produces generative cell by asymmetric division.

B. Pollen programme

5 Generative cell attached to pollen wall.

6 Generative cell rounds off within cytoplasm of vegetative cell;
OR
6″ Generative cell does not round off, but remains attached to pollen wall.

7 Generative cell divides by mitosis to form pair of sperm cells;
OR
7′ Generative cell divides to form pair of sperm cells which are linked together and with the vegetative nucleus forming a male germ unit;
OR
7″ No sperm cell division within maturing pollen grain.

8 Vegetative cell accumulates reserves, resulting in starch-rich pollen;
OR
8′ Starch hydrolysed to give lipid-rich pollen.

9 Germinal apertures with thickened intine layer;
OR
9′ Germinal apertures with wall ingrowths and membrane labyrinths;
OR
9″ No pre-formed germinal apertures present.

- fluorescence epi-illumination for FCR test of pollen quality and monitoring pollen developmental periods, both by the FCR test and use of DNA fluorochromes.

In addition, because of the small size of pollen, the greater resolving power of transmission electron microscopy may be required. These techniques need to be used with an appreciation of the requirements for valid developmental analysis. The adoption of evolutionary connotations for the characters of pollen ontogeny is premature, but will be dependent on careful analysis of developmental sequences using all these tools. It seems likely that some of the characters indicated with a prime symbol may prove to be synapomorphic after appropriate cladistic analyses. The characters of pollen ontogeny have particular potential in evolutionary studies of large genera, such as *Eucalyptus* and *Acacia*.

Acknowledgements

I thank the Australian Department of Education and Special Research Centres Program for financial support; Dr Chris Humphries (British Museum, Natural History, London) and Dr Cameron McConchie (now at the University of Heidelberg) for their kindness and scholarship without which parts of this article could not have been written; Dr Pauline Ladiges and Mr Andrew Drinnan for helpful discussions; Ms Terryn Hough for her skill in preparing Fig. 1, and Miss Debra Irvine as well as Mrs Susy Dreyer for their helpful secretarial assistance.

References

Alberch, P. (1985). Problems with the interpretation of developmental sequences. *Syst. Zool.* **34**, 46–58.

Beattie, A. J., Turnbull, C., Hough, T., Jobson, S. and Knox, R. B. (1984). The vulnerability of pollen and fungal spores to ant secretions: evidence and some evolutionary implications. *Am. J. Bot.* **72**, 606–614.

Blackmore, S. and Crane, P. R. (In press). The systematic implications of pollen and spore ontogeny. *In* "The Cladistics" (C. Humphreys, ed.). Academic Press, London.

Brink, R. A. and MacGillivray, J. A. (1924). Segregation for the waxy character in maize pollen and differential development of the male gametophyte. *Am. J. Bot.* **11**, 465–469.

Brummitt, R. K., Ferguson, J. K. and Poole, M. M. (1980). A unique and extraordinary pollen type in the genus *Crossandra* (Acanthaceae). *Pollen and Spores* **22**, 11–16.

Cass, D., Hough, T., McConchie, C. A. and Singh, M. B. (1986). Isolation of sperms from pollen of corn and oilseed rape. *Plant Physiol.* **130**, (Suppl.), 130.

Cruden, R. W. (1977). Pollen-ovule ratios: a conservative indication of breeding systems in flowering plants. *Evoltuion* **31**, 32–46.

Demerec, M. (1924). A case of pollen dimorphism in maize. *Am. J. Bot.* **11**, 461–464.

Ducker, S. C. and Knox, R. B. (1976). Submarine pollination in seagrasses. *Nature* **263**, 705–706.

Ducker, S. C., Pettitt, J. M. and Knox, R. B. (1978). Biology of Australian seagrasses: pollen development and submarine pollination in *Amphibolis antarctica* and *Thalassodendron ciliatum* (Cymodoceaceae). *Aust. J. Bot.* **26**, 265–285.

Dumas, C., Knox, R. B. and Gaude, T. (1984a). Pollen–pistil interactions: new concepts from electron microscopy and cytochemistry. *Int. Rev. Cytol.* **90**, 230–272.

Dumas, C., Knox, R. B., McConchie, C. and Russell, S. D. (1984b). Emerging physiological concepts in fertilization. *What's New in Plant Physiology* **15**, 17–20.

Dumas, C., Knox, R. B. and Gaude, T. (1985). The spatial association of the sperm cells and vegetative nucleus in the pollen grain of *Brassica. Protoplasma* **124**, 168–174.

Freeling, M. (1976). Introgenic recombination in maize: pollen analysis methods and the effect of parental Adh + isoalleles. *Genetics* **83**, 701–717.

Hennig, W. (1966). "Phylogenetic Systematics". University of Illinois Press, Urbana, USA.

Heslop-Harrison, J. (1968). Pollen wall development. *Science* **161**, 230–237.

Heslop-Harrison, J. and Heslop-Harrison, Y. (1970). Evolution of pollen viability by enzymatically induced fluorescence: intracellular hydrolysis of fluorescin diacetate. *Stain Technol.* **45**, 115–120.

Heslop-Harrison, J., Heslop-Harrison, Y. and Shivanna, K. R. (1984). The evolution of pollen quality, a further appraisal of the fluorochromatic (FCR) test procedure. *Theor. Appl. Genet.* **67**, 367–375.

Hough, T., Singh, M. B., Smart, J. J. and Knox, R. B. (1986). Immunofluorescent screening of monoclonal antibodies of animal and plant cells bound to polycarbonate membranes. *J. Immunol. Methods* **92**, 103–107.

Jensen, W. A. and Fisher, D. B. (1968). Cotton embryogenesis: the sperm. *Protoplasma* **65**, 277–286.

Kaul, V., Theunis, C. T., Palser, P., Knox, R. B. and Williams, E. G. (1987). Association of the generative cell and vegetative nucleus in pollen tubes of *Rhododendron. Ann. Bot.* **59**, 227–235.

Kenrick, J. and Knox, R. B. (1979). Pollen development and cytochemistry in some Australian species of *Acacia. Aust. J. Bot.* **27**, 413–427.

Kenrick, J. and Knox, R. B. (1982). Function of the polyad in reproduction of *Acacia. Ann. Bot.* **50**, 721–727.

Knox, R. B. (1984a). The pollen grain. *In* "Embryology of Plants" (B. M. Johri, ed.), pp. 197–271. Springer, Berlin, Heidelberg and New York.

Knox, R. B. (1984b). Pollen–pistil interaction. *In* "Encyclopaedia of Plant Physiology", Vol. 7 (H. F. Linkens and J. Heslop-Harrison, eds), pp. 508–608. Springer, Berlin, Heidelberg and New York.

Knox, R. B. and Heslop-Harrison, J. (1970). Direct demonstration of the low permeability of the angiosperm meiotic tetrad using a fluorogenic ester. *Z. Pflanzenphysiol.* **62**, 451–459.

Knox, R. B. and Kenrick, J. (1983). Polyad function in relation to the breeding system in *Acacia. In* "Pollen Biology" (D. Mulcahy and E. Ottaviano, eds), pp. 411–417. Elsevier-North Holland, Amsterdam.

Knox, R. B. and McConchie, C. (1986). Structure and function of compound pollen *In* "Pollen and Spores: Form and Function" (S. Blackmore and K. Ferguson, eds), pp. 265–282. Academic Press, London.

Maheshwari, P. (1950). "An Introduction to the Embryology of Angiosperms". McGraw-Hill, New York.

Mascarenhas, J. P., Willing, R. P. and Eisenberg, N. N. (1984). Pollen and gamete selection. *In* "Pollination '84" (E. G. Williams and R. B. Knox, eds), pp. 43–47. School of Botany, University of Melbourne.

McConchie, C. A. (1983). The reproductive biology of some Australian aquatic monocotyledons. Ph.D. Thesis, School of Botany, University of Melbourne.

McConchie, C. A. and Knox, R. B. (Submitted a). Pollen wall structure and development in the epihydrophilous *Lepilaena cylindrocarpa* (Zannichelliaceae). *Can. J. Bot.*

McConchie, C. A. and Knox, R. B. (Submitted b). Structure and development in the water-dispersed pollen of *Najas tenuifolia* (Najadaceae). *Am. J. Bot.*

McConchie, C. A., Knox, R. B. and Ducker, S. C. (1982). Ultrastructure and cytochemistry of the hydrophilous pollen of *Lepilaena* (Zannichelliaceae). *Micron* **13**, 339–340.

McConchie, C. A., Jobson, S. and Knox, R. B. (1985). Computer-assisted reconstruction of the male germ unit in pollen of *Brassica campestris*. *Protoplasma* **127**, 57–63.

McConchie, C. A., Hough, T. and Knox, R. B. (1987). Ultrastructural analysis of the sperm cells of mature pollen of maize, *Zea mays*. *Protoplasma* (in press).

McConchie, C. A., Russell, S. D., Dumas, C., Tuohy, M. and Knox, R. B. (1987). Quantitative cytology of the sperm cells of *Brassica campestris* and *Brassica oleracea*. *Planta* (in press).

McCoy, K. and Knox, R. B. (Submitted). The plasma membrane and generative cell organization in pollen of the Mimosoid legume *Acacia retinoides*. *Protoplasma*.

Metcalfe, C. R. (ed.) (1982). "Anatomy of the Monocotyledons", Vol. VII: Helobiae (Alismalidae). Clarendon Press, Oxford.

Mogensen, H. L. (1986). On the male germ unit in an angiosperm with bicellular pollen, *Hippeastrum vitatum*. *In* "Pollen Biotechnology and Ecology" (D. Mulcahy, G. Mulcahy and E. Ottaviano, eds), pp. 297–306. Springer, New York.

Mogensen, H. L. and Rusche, M. L. (1985). Quantitative analysis of barley sperm: occurrence and mechanism of cytoplasm and organelle reduction and the question of sperm dimorphism. *Protoplasma* **128**, 1–13.

Nelson, G. (1978). Ontogeny, phylogy, palaentology and the biogenetic law. *Syst. Zool.* **27**, 324–345.

Newman, J. V. (1933). Studies in the Australian Acacias. II. The life history of *Acacia baileyana* (FvM). Part I. Some ecological and vegetative features, spore production and chromosome number. *J. Linn. Soc. Bot.* **49**, 145–171.

Pettitt, J. M., McConchie, C. A., Ducker, S. C. and Knox, R. B. (1984). Reproduction in seagrasses: pollen wall morphogenetic in *Amphibolis antarctica* and wall structure in filiform grains. *Nordic J. Bot.* **4**, 199–216.

Russell, S. D. (1984). Ultrastructure of the sperm of *Plumbago zeylanica*. II. Quantitative cytology and three-dimensional organization. *Planta* **162**, 385–391.

Russell, S. D. (1985). Preferential fertilization in *Plumbago*: Ultrastructural evidence for gamete-level recognition in an angiosperm. *Proc. Natl Acad. Sci. U.S.A.* **82**, 6129–6132.

Russell, S. D. (1986). Isolation of sperm cells from the pollen of *Plumbago zeylanica*. *Plant Physiol.* **81**, 371–319.

Russell, S. D. and Cass, D. D. (1983). Unequal distribution of plastids and mitochondria during sperm cell formation in *Plumbago zeylanica*. *In* "Pollen: Biology and Implication for Plant Breeding" (D. L. Mulcahy and E. Ottaviano, eds), pp. 135–140. Elsevier, Amsterdam.

Singh, M. B. and Knox, R. B. (1984). Quantitative cytochemistry of β-galactosidase in normal (*Gal*) and deficient (*gal*) pollen of *Brassica campestris*: application of the indigogenic method. *Histochem. J.* **16**, 1273–1296.

Singh, M. B. and Knox, R. B. (1985). A gene controlling β-galactosidase deficiency in pollen of oilseed rape. *J. Hered.* **76**, 199–201.

Singh, M. B., O'Neill, P. and Knox, R. B. (1985). Initiation of postmeiotic β-galactosidase synthesis during microsporogenesis in oilseed rape. *Plant Physiol.* **77**, 225–228.

Smart, J. J., Tuddenham, W. G. and Knox, R. B. (1979). Aerobiology of grass pollen in the city atmosphere of Melbourne: effects of weather parameters and pollen sources. *Aust. J. Bot.* **27**, 333–342.

Stern, H. (1966). The regulation of cell division. *Ann. Rev. Plant Physiol.* **17**, 345–378.

Stinson, J. and Mascarenhas, J. P. (1985). The onset of alcohol dehydrogenase synthesis during microsporogenesis in maize. *Plant Physiol.* **77**, 222–224.

Vergne, P., Delvallee, I. and Dumas, C. (In press). Rapid assessment of microspore and pollen development stages in wheat (*Triticum aestivum* L.) and maize (*Zea mays* L.) using a DNA-specific fluorochrome. *J. Cell Sci.*.

Vierssen, W. van, Van Wijk, R. J. and Vander Zee, J. R. (1982). On the pollination mechanisms of some eurysaline Potamogetonaceae. *Aquat. Bot.* **14**, 339–347.

Vithanage, H. I. M. V. and Knox, R. B. (1980). Periodicity of pollen development and quantitative cytochemistry of exine and intine enzymes in the grasses *Lolium perenne* and *Phalaris tuberosa*. *Ann. Bot.* **45**, 131–142.

Wiley, E. D. (1981). "Phylogenetics: The Theory and Practice and Phylogenetic Analysis". Wiley, New York.

Willemse, N. I. R. M. and Reznickova, S. A. (1980). Formation of pollen in the anther of *Lilium*. I. Development of the pollen wall. *Acta Bot. Neerl.* **29**, 127–140.

Willson, M. (1983). "Plant Reproductive Ecology". Wiley, Toronto and New York.

Wilms, J. H. and van Aelst, A. C. (1983). Ultrastructure of spinach sperm cells in mature pollen. *In* "Fertilization and Embryogenesis in Ovulated Plants", pp. 105–111. Veda, Bratislava.

3 Gene Flow and Distribution of Genetic Variation in Plant Populations

JAMES L. HAMRICK

Departments of Botany and Genetics, University of Georgia, Athens, Georgia, U.S.A.

I. Introduction

The movement of genes among populations or population subdivisions has two distinct but interrelated effects on the genetic structure of plant populations. High rates of gene flow increase effective population sizes (N_e) and reduce genetic differentiation by drift or natural selection (Jain and Bradshaw, 1966). Lower levels of gene flow affect genetic variation within populations by introducing novel alleles and genotypes, but may have relatively little effect on the frequencies of common alleles (Antonovics, 1968). Therefore, while gene flow may often be too low to have profound effects on population differentiation, it can have an important role to play in the evolution of populations by introducing and maintaining genetic variation within populations.

The exact evolutionary role played by gene flow has been the subject of much debate and may depend on complex interactions among several evolutionary parameters. In populations or species subject to intense selection pressures or those that are geographically isolated, gene flow may play a minor role. Conversely, in species that experience lower selection pressures or whose populations are continuous relative to the species' ability to migrate, gene flow could play an important role in shaping the distribution of genetic variation. Wright (1948) has shown that if the value $4Nm$ (where N is the effective population size and m is the rate of gene flow; or Nm is the number of individuals migrating into a population per generation) is less than 1, genetic drift will play a major

Differentiation Patterns in Higher Plants
ISBN: 0-12-709425-3

role in determining the distribution of genetic variation among populations. On the other hand, if $4Nm$ is much greater than 1 genetic drift will play a minor role. If neutral alleles are assumed Wright (1951) has shown that the commonly used measure of genetic differentiation F_{ST} is equal to $1/(4Nm + 1)$. As a result genetic differentiation will be greatly reduced even with low levels of gene flow (e.g. $Nm = 1$). Estimates of the magnitude of gene flow are therefore essential for accurate predictions of population differentiation.

The magnitude of gene flow has been the subject of much discussion but little direct measurement. Earlier writers assumed that gene flow was high and that it served to "maintain the integrity of species" (Mayr, 1963). This viewpoint was challenged by Ehrlich and Raven (1969), who concluded that gene flow was limited for most species and that genetic similarities shared by spatially isolated populations were due to natural selection. Levin and Kerster (1974), after reviewing the available plant literature, concluded that gene flow in most plant species was less than 1% among populations or population subdivisions separated by a few hundred metres and at least two orders of magnitude less among populations 1.5 kilometres or more apart. In predominantly self-fertilized species without special mechanisms for long distance seed dispersal, effective gene flow among populations was predicted to be at levels approaching mutation rates (i.e., 10^{-5} to 10^{-6}). More recent evidence, however, has caused Levin (1983) and others (Schaal, 1980) to revise their estimates upwards.

A glaring deficiency of these discussions is the scarcity of direct measures of gene flow. In plants this problem is compounded by the complexity of the processes involved with gene movement. Gene movement in plants is usually accomplished by the movement of the male gamete (pollen) or as an immature embryo (the seed or fruit). Also, many plant species are capable of self-fertilization to some degree. Thus, accurate estimates of gene flow in plant populations require measures of the mating system (percentage of selfing/outcrossing) as well as pollen and seed movement.

II. Gene Flow Measurements

A. *Traditional procedures*

Several procedures are commonly used to measure gene flow in plants. Juvenile frequencies or means can be compared with those of the adult generation. This appears to be a straightforward approach, but in practice it is often of limited use. Although it is relatively simple to calculate genotype frequencies at different stages of the plant life cycle, changes in genotype frequencies may be due to any of several evolutionary factors, especially non-random mating and selection. Therefore, before gene flow can be estimated, direct measures of the breeding system and the direction and intensity of selection must be available. Measures of selection intensity may be difficult to obtain since selection can act in several ways to change genotype frequencies. Certain genotypes may produce more pollen than other genotypes, or selection can act at any of the life cycle stages between pollen production and the juvenile plant. Unless these parameters have been accurately quantified, it will be impossible to estimate the effects of gene flow on the genetic composition of populations.

Most workers attempt to measure gene flow by following the movement of seed or pollen dispersal vectors, the movement of dispersal units (pollen or seeds) or the movement of marker genes. Studies of dispersal agents have the advantage that movements can be followed from several plants within the population. As a result the pollination or seed dispersal ecology of individuals located in a variety of densities, patch sizes or positions within patches (edge vs. centre) can be examined. The validity of this approach as a measure of gene flow is, however, based on several assumptions. First, movements of pollen or seed dispersal agents are assumed to represent patterns of pollen or seed deposition. There is an increasing amount of evidence that pollinator movements under-represent actual gene movement due to high pollen carryover (Lertzman and Gass, 1983; Levin, 1983; Schaal, 1980). Secondary movement has also been documented for seeds and has the potential to greatly increase gene flow (Beattie and Culver, 1979). Second, this approach provides no information on the mating system. Plants that experience high levels of selfing coupled with long distance movement of the released pollen may have less gene flow than predominantly outcrossing species whose pollen is carried short distances. Finally, the movements of dispersal agents give no information on the success of the pollen grain in fertilization or about the successful establishment of a seed.

Following the actual movement of pollen or seed rather than their vectors is more difficult, but can be accomplished by marking pollen or seed with dyes or radioactive tracers or by measuring pollen or seed deposition around isolated individuals. Since actual pollen or seed deposition patterns are described, problems caused by pollen carryover are circumvented and some of the problems of estimating rates of selfing/outcrossing are reduced. The major drawback of this procedure is that it provides no information on fertilization or seedling success. Also for many species, especially trees, it may be difficult or impossible to mark pollen or seed.

Many of the problems discussed above do not arise if the movement of unique alleles is followed directly. A serious problem, however, is the difficulty of finding individuals in natural populations with a unique allele. At best only a few source plants will be available and, as a result, generalizations concerning gene movement in populations or species must be made from the pollen and seed deposition patterns of one or a few individuals. Consequently, genetic markers have most often been used in artificially established populations (e.g., Batemen, 1947; Handel, 1982, 1983a; Smyth and Hamrick, in press). Here too, there are several problems for it is often difficult to mimic the characteristics (density, dispersion, size structure, pollinators, etc.) of natural populations.

In summary, although the procedures which have commonly been used to estimate gene movement have provided insights into the evolutionary mechanisms of plant populations, they have not consistently provided the level of detail needed to make quantitative predictions concerning population differentiation.

B. New approaches

Recently two new approaches have been developed to measure gene movement. The first procedure, paternity analysis, provides a detailed description of the breeding structure of populations. Paternity analyses are based on using polymorphic genetic loci to identify the most probable father (or parents in the case of a seed or seedling) from a set of candidates (Meagher and Thompson, 1986; Hamrick and Schnabel, 1985). The resolution of the procedure is dependent on the number of polymorphic loci, the number of alleles per locus, the frequency of those alleles, and the number of potential parents. Once the sire of each progeny has been identified patterns of pollen movement within the study area can be determined. If several maternal plants are studied, pollen dispersal away from paternal individuals can be described (Meagher, in press) and the proportion of effective pollen originating outside the study area can be determined. Also, by carefully selecting the location of monitored plants or by experimental manipulations the effects of density, dispersion and the location of individuals (edge vs. centre) on patterns of incoming and outgoing pollen can be described (Ellstrand and Marshall, 1985).

Slatkin (1981, 1985a) has developed a novel approach for the indirect estimation of gene flow. Allele frequency distributions among populations are used to estimate the average number of migrants exchanged between local populations (Nm; where N = population size and m is the fraction of N replaced by immigrants). The natural logarithm of Nm is approximately linearly related to the natural logarithm of the mean frequency of private alleles (i.e., alleles that appear in only one population; Slatkin, 1985a). This approach has great appeal since it can be applied directly to the electrophoretic data available for large numbers of plant species. The procedure, however, has not been compared with more direct measures of gene flow to determine whether it provides a dependable quantification of gene flow or is more useful as a comparative statistic.

In the following sections the Slatkin (1985a) method is used to calculate estimates of gene flow for several groups of plant species. These estimates will be compared to genetic diversity measures (G_{ST}) which have been calculated from the same data sets. It will be shown that plant species vary widely in their ability to transport genes among populations and that these differences have profound effects on the spatial distribution of genetic variation.

III. Genetic Structure

Genetic structure can be defined as the nonrandom distribution of alleles or genotypes in space or time (Loveless and Hamrick, 1984). The most informative and most widely employed statistics for the description of genetic structure are Wright's (1951) "*F*" statistics and the genetic diversity statistics of Nei (1973). "*F*" statistics are based on the deviations of genotype frequencies from frequencies expected under random mating. The total deviation (F_{IT}) is partitioned into deviations due to variation in allele frequencies among population subdivisions (F_{ST}), the Wahlund effect, and deviations from expectations within population subdivisions (F_{IS}). These parameters are related as:

$$F_{IT} = F_{ST} + (1 - F_{ST})F_{IS}$$

Nei's genetic diversity statistics partition the total genetic diversity (H_T) into the mean diversity within subpopulations (H_S) and the gene diversity among population subdivisions (D_{ST}). Thus,

$$H_T = H_S + D_{ST}$$

The proportion of gene diversity among populations relative to the total gene diversity (D_{ST}/H_T) is defined as G_{ST}. Nei (1977) has shown that his G_{ST} statistic is equivalent to a multi-allelic F_{ST}. These two statistics are, therefore, often used interchangeably.

Plant populations do not normally consist of randomly arranged assemblages of genotypes. Significant genetic heterogeneity has been shown to exist among geographically distinct populations, among subdivisions of local populations, or even among the progeny of individual plants (Allard, 1975; Hamrick and Holden, 1979; Linhart *et al.*, 1981; Schaal, 1975). The genetic structure seen in natural populations is assumed to have resulted from the action of mutation, migration, selection and drift. In a recent review, Loveless and Hamrick (1984) demonstrated that the distribution of allozyme variation was closely associated with several characteristics of the species. Measures of intrapopulation genetic diversity (H_S) were consistently higher in species that combined character states of the breeding system, life cycle, timing of reproduction, successional stage, and geographic range that encouraged gene flow. Similarly, measures of interpopulation differentiation (G_{ST}) were consistently lower in species that combined characteristics that promoted gene flow: outcrossed, wind-pollinated, long-lived, polycarpic species of the later stages of succession. Seed dispersal mechanisms were not good predictors of genetic structure. At the extremes, species whose seeds are dispersed by the wind demonstrated little genetic differentiation while gravity dispersed species had high G_{ST} values (Table 1). The major difficulty arose in predicting the effect of animal dispersal on genetic structure. Species with animal-ingested and animal-attached seeds had high G_{ST} values; contrary to predictions.

In order to compare the relationship between gene flow and genetic differentiation among populations, estimates of *Nm* have been obtained for groups of plant

TABLE 1

Estimates of *Nm* based on Slatkin's (1985a) procedure of using rare alleles as indicators of gene flow. Values were corrected for mean sample sizes. Mean G_{ST} values are also given for each category (from Hamrick and Griswold, in press)

	Number of studies	$\bar{p}(1)$	*Nm*	G_{ST}
Breeding system				
Selfed	19	0.316	0.065	0.485
Mixed – animal	9	0.200	0.161	0.256
Outcrossed – animal	16	0.089	0.801	0.178
Outcrossed – wind	14	0.034	5.380	0.050
Seed disperal mechanism				
Gravity				
Large	10	0.200	0.161	0.285
Small	15	0.315	0.065	0.397
Animal – attached	9	0.148	0.292	0.346
Animal – ingested	4	0.194	0.171	0.287
Winged or plumose	20	0.038	4.313	0.099

species with different breeding systems and seed dispersal mechanisms (Hamrick and Griswold, in press). Mean estimates of *Nm* ranged from 0.065 for selfing plants to 5.38 for outcrossed wind-pollinated species (Table 1). Mixed-mating species and outcrossed animal-pollinated species had intermediate values of *Nm*. A similar pattern was seen for the groups based on seed dispersal mechanisms. Small-seeded, gravity-dispersed species had the lowest values while winged or plumose seeded species had the highest.

Mean values of *Nm* can be compared to measures of interpopulation divergence (G_{ST}). It might be argued that this comparison is circular since the same data are used in both analyses. However, estimates of *Nm* are based on the distribution of private alleles while G_{ST} values are largely dependent on the distribution of common alleles. The values in Table 1 indicate that the G_{ST} values closely parallel estimates of *Nm*.

One of the problems with averaging over several species is that individual studies often represent different sampling designs. With Slatkin's procedure estimates of *Nm* based on geographically widespread populations should be smaller than those based on geographically restricted populations or population subdivisions. Thus, the range in *Nm* values should be much larger if individual species are compared. In Table 2 estimates of *Nm* range from 0.050 for *Phlox cuspidata* to 37.8 for *Pinus ponderosa*. Once again, measures of *Nm* are closely associated with the G_{ST} values of each species. A Spearman's rank correlation analysis (Steel and Torrie, 1960) based on these 15 species was highly significant (r_s = 0.889; $P < 0.001$). It therefore appears that the genetic mobility of a species has a direct influence on the spatial distribution of allozyme variation.

TABLE 2

Values of $\bar{p}(1)$, Nm and G_{ST} for representative species classified by their breeding system. Geographic range is the spatial area represented in the study (R = regional; L = local; H = highly local)

Species type	Species	Geographic range	No. private alleles	$\bar{p}(1)$	Nm	G_{ST}	Source
Selfing	*Chenopodium praticola*	R	3	0.383	0.090	0.477	Crawford and Wilson (1979)
	Triticum dioccoides	L	13	0.260	0.082	0.449	Nevo *et al.* (1982)
	Hordeum jubatum	L	4	0.078	0.477	0.259	Shumaker and Babbel (1980)
	Hordeum spontaneum	L	12	0.263	0.061	0.360	Nevo *et al.* (1979)
Mixed – mating	*Eucalyptus caesia*	R	5	0.080	0.836	0.315	Moran and Hopper (1983)
	Lupinus subcarnosus	L	3	0.107	0.757	0.383	Babbel and Selander (1974)
	Phlox cuspidata	L	2	0.375	0.050	0.455	Levin (1975)
Outcrossed – animal	*Coreopsis cyclocarpa* var. *pinnatisecta*	R	9	0.157	0.244	0.186	Crawford and Bayer (1981)
	Mimulus guttatus	H	5	0.046	0.605	0.213	McClure (1973)
	Stephanomeria exigua	L	5	0.053	1.47	0.077	Gottlieb (1975)
	Swartzia simplex var. *ochnacea*	H	12	0.014	13.9	0.055	Hamrick and Loveless (in prep.)
Outcrossed – wind	*Pinus contorta*	H	4	0.010	36.4	0.016	Hamrick and Smith (in prep.)
	P. ponderosa	L	4	0.004	37.8	0.015	Hamrick *et al.* (in prep.)
	P. rigida	R	8	0.013	17.2	0.023	Guries and Ledig (1982)
	Pseudotsuga menziesii	H	7	0.058	5.3	0.068	El-Kassaby and Sziklai (1982)

There are, however, some interesting inconsistencies among these 15 species (Table 2). First, contrary to predictions, the spatial distribution of populations has little relationship with estimates of *Nm*. Second, within each group of species at least one has a conspicuously high or low estimate of *Nm*. Among the predominantly selfing species, *Hordeum jubatum* (Shumaker and Babbel, 1980) had a much higher estimate of *Nm* and a lower G_{ST} value (Table 2). *H. jubatum* is a species of open grasslands and weedy roadsides throughout the western plains and the intermountain region. The seeds of *H. jubatum* are adapted for transport by attachment to animal fur and can be carried long distances. Furthermore, its long awns can become intertwined to form a cluster of seeds that may be blown long distances by the strong winds that characterize much of its native range. Thus, the potential for long-distance seed dispersal may lead to more gene flow than its breeding system would indicate. The low estimated values of *Nm* obtained for *Phlox cuspidata* (Levin, 1975) and *Coreopsis cyclocarpa* var. *pinnatisecta* (Crawford and Bayer, 1981) are more difficult to explain. Levin (1975) states that *P. cuspidata* is a facultatively selfing species. This analysis (Table 2), while based on only two private alleles, indicates that *P. cuspidata* may be predominantly selfing. *C. cyclocarpa* var. *pinnatisecta* may also be more highly selfing than its classification indicates. Estimates of *Nm* indicate that it would better fit into the mixed-mating group. On the other hand, G_{ST} values for this species are not appreciably higher than the mean of the outcrossed group (Table 1). Finally, the tropical tree *Swartzia simplex* var. *ochnacea* (Hamrick and Loveless, in prep.) has a higher estimate of *Nm* and lower estimate of G_{ST} than its classification might indicate. *Swartzia* is predominantly pollinated by large, strong-flying bees which Janzen (1971) has shown are capable to flying several hundred metres while foraging. The local distribution of the sites sampled coupled with the distinct possibility of pollen carryover by these bees could account for the high rates of gene flow observed.

IV. Patterns of Gene Flow

Population geneticists have developed several theoretical models to describe gene flow. The most commonly used models are island, stepping-stone, and isolation-by-distance (Fig. 1). With the island model a population is subdivided into several panmictic units, each receiving a proportion of the migrants drawn at random from the entire population. Since each subpopulation is equally likely to send individuals to every other subpopulation the island model represents the extreme in long-distance gene flow. With the stepping-stone model gene flow occurs only between adjacent populations. In a single dimension such as a linearly distributed group of populations, gene flow is between a population and its two immediate neighbours. In groups of populations with two distinct dimensions gene flow is assumed to occur between the population and its four or six immediate neighbours depending on whether a rectangular or hexagonal lattice is assumed (Slatkin, 1985b).

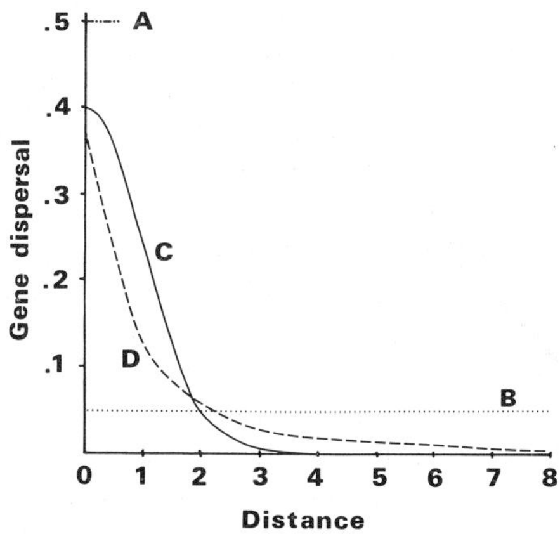

FIG. 1 Patterns of gene dispersal $F(x)$ as a function of distance. (A) Stepping-stone model: $F(x) = 0.50$; (B) island model: $F(x) = 0.50$; (C) normal distribution: $F(x) = \exp(-x)$; (D) leptokurtic distribution: $F(x) = \exp(-x^{1/2})/x$ (after Jain and Bradshaw, 1966).

The stepping-stone model represents the extreme in short-distance gene flow. In a more or less continuously distributed species gene flow decreases as spatial separation increases due to the limited dispersal of gametes or zygotes. This pattern has been termed isolation-by-distance. Typically, two distributions of gene dispersal are assumed; a normal distribution and a leptokurtic distribution. These two distributions differ in that the leptokurtic distribution has more short-distance and more long-distance dispersal than does the normal distribution (Fig. 1). The leptokurtic distribution has a larger variance in dispersal distances and results in larger effective neighbourhood sizes, greater gene flow and less population differentiation (Jain and Bradshaw, 1966).

Direct measures of pollen or seed dispersal indicate that the isolation-by-distance model best fits the patterns observed (Levin and Kerster, 1968; Muller, 1977; Smyth and Hamrick, in press; Webb and Bawa, 1983). This evidence, which is based primarily on pollinator movements or by following the distribution of unique alleles in artificial populations, indicates that most pollen or seed are deposited near their source but that occasional long-distance moves occur. More recent studies that have simultaneously used pollinator movements and marker genes indicate that estimates based on pollinator moves may underestimate pollen movements by at least 100% (Levin, 1983; Schaal, 1980; C. Fenster, pers. comm.). This result is presumably due to the occurrence of considerable pollen carryover. The research of Thomson and Plowright (1980), Price and Waser (1982), and Lertzmann and Gass (1983), who have examined pollen carryover directly, support this conclusion. These studies demonstrated that pollen from marked individuals may continue to be deposited after as many as 50 subsequent moves. Furthermore, the length and height of the tail of the leptokurtic distribution is often poorly quantified since most studies are carried out in small, well-defined populations. If the tails of these distributions extend far beyond the boundaries of the study area gene flow may be much greater than has been indicated (Slatkin, 1985b).

Published descriptions of pollen and seed dispersal usually present mean values that have been determined by examining several individuals. Little attention is typically paid to variation in pollen or seed deposition at locations at equal distances from the source plants. When actual data is examined (Fig. 2) there is usually considerable heterogeneity among sites for the number of marker genes deposited. Thus, while the overall pattern of pollen or seed dispersal with increasing distance may produce a smooth leptokurtic distribution, the actual pattern may be quite patchy. There is also the possibility that the level of clumping may increase with distance (Fig. 2). Since each patch of marked pollen or seeds may represent the action of a single dispersal agent, there is a strong possibility that the pollen or seed within each patch are genetically related. Such distributions should have a marked influence on the development of genetic structure.

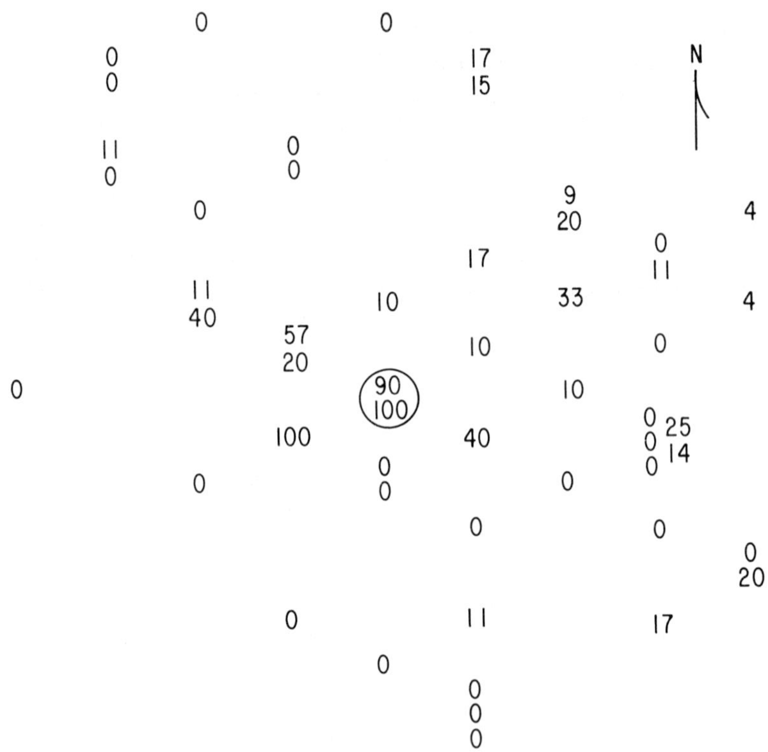

FIG. 2 The distribution of a unique marker allele (PGI-S) in an experimental population of *Carduus nutans*. The numbers represent the proportion of heterozygous progeny. Two or more numbers at a location represent values for individual flowering heads. The source plant is circled. Note the heterogeneity in marker allele frequencies among plants at equivalent distances and among heads within a plant (from Smyth and Hamrick, 1987).

To my knowledge no one has formally modelled the effects of the patchy distribution of pollen and seed on the development and maintenance of genetic structure. Turner *et al.* (1982) have simulated the effects of nearest-neighbour mating and localized seed dispersal and have shown that considerable genetic heterogeneity develops. Hamrick and Loveless (1986) developed a verbal model in which they compared the effects of various breeding systems, seed dispersal mechanisms and selection on the subsequent development of genetic structure within and among populations. At one extreme they assumed that mating was among nearest neighbours and that seeds were dispersed in clumps of related individuals. The nearest neighbour matings increased the probability that adjacent seedlings had both of their parents in common. When combined with the clumped distribution of seeds, considerable genetic structure should develop among the patches of seedlings or among populations established after long distance dispersal. Selective mortality would act to maintain this structure, especially if microhabitat selection favours certain full-sib families.

Conversely, species with more nearly random mating within their populations (and subsequently more gene flow among populations) combined with a wave-like leptokurtic distribution of seed around the parent plants should maintain much less genetic structure. The genetic diversity of the progeny of individual maternal plants should be high and the genetic relatedness of the progeny low. Subsequent mortality should act to decrease genetic structure unless post-dispersal mortality selects for particular half-sib families in different microhabitats.

It appears therefore that descriptions of gene flow that focus on the average number of migrants at a given distance from the source may lead to underestimates of the potential for the development of genetic structure. Certainly more attention needs to be paid to the actual patterns of gene movement if the causes of genetic differentiation are to be understood.

V. Conclusions

The influence of gene flow on the structure and evolution of natural plant populations remains controversial. Much of the discussion has centred around the magnitude of gene flow among populations. Indirect estimates based on allele frequency distributions produce higher estimates of gene flow than direct measures based on following dispersal units or marker genes. These discrepancies may be accounted for by the limitations of most experimental designs for quantitatively measuring truly long-distance dispersal. Yet, these relatively rare, long-distance events are the most effective in increasing neighbourhood sizes and decreasing population differentiation. Procedures such as paternity analyses which give detailed descriptions of the breeding structure of populations should provide additional insights. The few studies available that have used multi-locus techniques to study gene flow indicate that other estimation procedures may underestimate the magnitude of gene flow for some plant species. Friedman and Adams (1985) found that as much as 40% of the pollen successfully fertilizing ovules in a loblolly pine (*Pinus taeda*) seed orchard had come from wild populations located 400 m from the orchard. Studies by Ellstrand and Marshall (1985) of isolated populations of the insect-pollinated wild radish, *Raphanus sativus*, indicate that rates of gene flow of 5 – 10% may be found between populations separated by as much as a kilometre.

Both pollen and seed may be dispersed in a highly clumped fashion and these patches may consist of genetically related individuals. If the patchy distribution of pollen and seed is a general phenomenon in plants it could have a major influence on the development of genetic heterogeneity within and among populations. More direct data are needed on the actual rates and patterns of gene flow. We have not paid enough attention to the actual distribution of dispersed genes or to the effects of density, dispersion or temporal variation on rates of gene flow. There is no reason to believe that gene flow should be constant among populations with different structures (Handel, 1983b), nor that gene flow is constant from year to year. In fact, the movement of genes among populations may be characterized by long periods of low gene flow punctuated by short periods when gene flow is much higher (Slatkin, 1985b). If such a pattern is characteristic of many species it might help to explain the differences which seem to occur between the direct and indirect measures of gene flow.

Gene flow rates vary greatly among plant species. Some species, selfing plants in particular, experience low levels of gene flow and as a result, have small neighbourhood sizes, which in turn leads to genetic differentiation. Wind-pollinated species such as conifers, on the other hand, have little genetic differentiation among their populations for allozyme traits and, as expected, estimates of gene flow for these species are quite high. The significant differences seen among populations of such species for morphological or physiological traits must be due to the action of rather intense selection, since gene flow should have equal effects on all genetically controlled traits.

Finally, if nothing else, this discussion should discourage the lumping of all plant species into a single category characterized by low gene flow and high genetic diversity among populations. The data presented indicate that gene flow among plant populations can be lower than that found among populations of the relatively immobile salamander, *Batrachoseps campi* ($Nm = 0.16$; Slatkin, 1985a), or as high as that measured for the marine bivalve *Mytilis edulis* ($Nm = 42$; Slatkin, 1985a) which has pelagic larvae. Certainly it cannot be said that plants in general lack genetic mobility.

Acknowledgements

I wish to acknowledge the helpful comments of M. D. Loveless. I am also indebted to her for the preparation of the figures. Portions of this research were supported by National Science Foundation Grants Nos. BSR8206946 and BSR8320529.

References

Allard, R. W. (1975). The mating system and microevolution. *Genetics* **79** (Suppl.), 115–126.

Antonovics, J. (1968). Evolution in closely adjacent plant populations. VI. Manifold effects of gene flow. *Heredity* **23**, 507–524.

Babbel, G. R. and Selander, R. K. (1974). Genetic variability in edaphically restricted and widespread plant species. *Evolution* **28**, 619–630.

Bateman, A. J. (1947). Contamination in seed crops. III. Relation with isolation distance. *Heredity* **1**, 303–306.

Beattie, A. J. and Culver, D. C. (1979). Neighborhood size in *Viola. Evolution* **33**, 1226–1229.

Crawford, D. J. and Bayer, R. J. (1981). Allozyme divergence in *Coreopsis cyclocarpa* (Compositae). *Syst. Bot.* **6**, 373–379.

Crawford, D. J. and Wilson, H. D. (1979). Allozyme variation in several closely related diploid species of *Chenopodium* of the western United States. *Am. J. Bot.* **66**, 237–244.

Ehrlich, P. R. and Raven, P. H. (1969). Differentiation of populations. *Science* **165**, 1228–1231.

El-Kassaby, U. A. and Sziklai, O. (1982). Genetic variation of allozyme and quantitative traits in a selected Douglas-Fir (*Pseudotsuga menziesii* var. *mensiezii* (Mirb.) Franco) population. *For. Ecol. Manag.* **4**, 115–126.

Ellstrand, N. C. and Marshall, D. L. (1985). Interpopulation gene flow by pollen in wild radish, *Raphanus sativus. Am. Nat.* **126**, 606–616.

Friedman, S. T. and Adams, W. T. (1985). Estimation of gene flow into two seed orchards of loblolly pine (*Pinus taeda* L.). *Theor. Appl. Genet.* **69**, 609–615.

Gottlieb, L. D. (1975). Allelic diversity in the outcrossing annual plant *Stephanomeria exigua* ssp. *carotifera* (Compositae). *Evolution* **29**, 213–225.

Guries, R. P. and Ledig, F. T. (1982). Genetic diversity and population structure in pitch pine (*Pinus rigida* Mill.). *Evolution* **36**, 387–402.

Hamrick, J. L. and Griswold, G. B. (In press). Association between Slatkin's measure of gene flow and the dispersal ability of plant species. *Am. Nat.*

Hamrick, J. L. and Holden, R. L. (1979). Influence of microhabitat heterogeneity on gene frequency distribution and gametic phase disequilibrium in *Avena barbata. Evolution* **33**, 521–533.

Hamrick, J. L. and Loveless, M. D. (1986). The influence of seed dispersal mechanisms on the genetic structure of plant populations. *In* "Frugivores and Seed Dispersal" (A. Estrada and T. H. Fleming, eds), pp. 211–223. W. Junk, Dordrecht.

Hamrick, J. L. and Loveless, M. D. (In prep.). The distribution of genetic variation in a tropical tree, *Swartzia simplex* var. *ochnacea.*

Hamrick, J. L. and Schnabel, A. (1985). Understanding the genetic structure of plant populations. Some old problems and a new approach. *In* "Population Genetics in Forestry", Lecture Notes in Biomathematics 60 (H. R. Gregorius, ed.), pp. 50–70. Springer, Berlin.

Hamrick, J. L. and Smith, C. C. (In prep.). Gene flow and genetic diversity among local populations of lodgepole pine.

Hamrick, J. L., Blanton, H. M. and Hamrick, K. J. (In prep.). The genetic structure of geographically marginal populations of ponderosa pine.

Handel, S. N. (1982). Dynamics of gene flow in an experimental population of *Cucumis melo* (Cucurbitaceae). *Am. J. Bot.* **69**, 1538–1546.

Handel, S. N. (1983a). Contrasting gene flow patterns and genetic subdivision in adjacent populations of *Cucumis sativus* (Cucurbitaceae). *Evolution* **37**, 760–771.

Handel, S. N. (1983b). Pollination ecology, plant population structure and gene flow. *In* "Population Ecology" (L. Real, ed.), pp. 163–212. Academic Press, New York.

Jain, S. K. and Bradshaw, A. D. (1966). Evolutionary divergence among adjacent plant populations. I. The evidence and its theoretical analysis. *Heredity* **21**, 407–441.

Janzen, D. H. (1971). Euglossine bees as long-distance pollinators of tropical plants. *Science* **171**, 203–205.

Lertzman, K. P. and Gass, C. L. (1983). Alternative models of pollen transfer. *In* "Handbook of Experimental Pollination Biology" (C. E. Jones and R. J. Little, Jr., eds), pp. 474–489. Van Nostrand Reinhold, New York.

Levin, D. A. (1975). Interspecific hybridization, heterozygosity and gene exchange in *Phlox*. *Evolution* **29**, 37–51.

Levin, D. A. (1983). Plant parentage: an alternative view of the breeding structure of populations. *In* "Population Biology: Retrospect and Prospect" (C. E. King and P. S. Dawson, eds), pp. 171–188. Columbia University Press, New York.

Levin, D. A. and Kerster, H. W. (1968). Local gene dispersal in *Phlox*. *Evolution* **22**, 130–139.

Levin, D. A. and Kerster, H. W. (1974). Gene flow in seed plants. *Evol. Biol.* **7**, 139–220.

Linhart, Y. B., Mitton, J. B., Sturgeon, K. B. and Davis, M. L. (1981). Genetic variation in space and time in a population of Ponderosa pine, *Pinus ponderosa*. *Heredity* **46**, 407–426.

Loveless, M. D. and Hamrick, J. L. (1984). Ecological determinants of genetic structure in plant populations. *Ann. Rev. Ecol. Syst.* **15**, 65–96.

Mayr, E. (1963). "Animal Species and Evolution". Harvard University Press, Cambridge, MA.

McClure, S. (1973). Allozyme variability in natural population of the yellow monkeyflower, *Mimulus guttatus*, located in the North Yuba River drainage. Ph.D. Dissertation, University of California, Berkeley, CA.

Meagher, T. R. (In press). Analysis of sibship structure in a natural population of *Chamaelirium luteum* (Liliaceae). *Am. Nat.*

Meagher, T. R. and Thompson, E. T. (1986). The relationship between single parent and parent pair genetic likelihoods in genealogy reconstruction. *Theor. Pop. Biol.* **29**, 87–106.

Moran, G. F. and Hopper, S. D. (1983). Genetic diversity and the insular population structure of the rare granite rock species, *Eucalyptus caesia* Benth. *Aust. J. Bot.* **31**, 161–172.

Muller, G. (1977). Cross-fertilization in a conifer stand inferred from enzyme gene-markers in seeds. *Silv. Genet.* **26**, 223–226.

Nei, M. (1973). Analysis of gene diversity in subdivided populations. *Proc. Natl Acad. Sci. U.S.A.* **70**, 3321–3323.

Nei, M. (1977). F-statistics and analysis of gene diversity in subdivided populations. *Ann. Hum. Genet.* **41**, 225–233.

Nevo, E., Zohary, D., Brown, A. H. D. and Haber, M. (1979). Genetic diversity and environmental associations of wild barley, *Hordeum spontaneum*, in Israel. *Evolution* **33**, 815–833.

Nevo, E., Golenberg, E., Beiles, A., Brown, A. H. D. and Zohary, D. (1982). Genetic diversity and environmental associations of wild wheat, *Triticum dicoccoides*, in Israel. *Theor. Appl. Genet.* **62**, 241–254.

Price, M. V. and Waser, N. M. (1982). Experimental studies of pollen carryover: hummingbirds and *Ipomopsis aggregata*. *Oecologia* **54**, 353–358.

Schaal, B. A. (1975). Population structure and local differentiation in *Liatris cylindracea*. *Am. Nat.* **109**, 511–528.

Schaal, B. A. (1980). Measurement of gene flow in *Lupinus texensis. Nature* **284**, 450–451.
Shumaker, K. M. and Babbel, G. R. (1980). Pattern of allozyme similarity in ecologically central and marginal populations of *Hordeum jubatum* in Utah. *Evolution* **34**, 110–116.
Slatkin, M. (1981). Estimating levels of gene flow in natural populations. *Genetics* **99**, 323–335.
Slatkin, M. (1985a). Rare alleles as indicators of gene flow. *Evolution* **39**, 53–65.
Slatkin, M. (1985b). Gene flow in natural populations. *Ann. Rev. Ecol. Syst.* **16**, 393–430.
Smyth, C. A. and Hamrick, J. L. (1987). Realized gene flow via pollen in artificial populations of *Carduus nutans. Evolution.* **41**, 613–619.
Steel, R. G. D. and Torrie, J. H. (1960). "Principles and Procedures of Statistics". McGraw-Hill, New York.
Thomson, J. D. and Plowright, R. C. (1980). Pollen carryover, nectar rewards, and pollinator behavior with special reference to *Diervilla lonicera. Oecologia* **46**, 68–74.
Turner, M. E., Stephens, J. C. and Anderson, W. W. (1982). Homozygosity and patch structure in plant populations as a result of nearest-neighbor pollination. *Proc. Natl Acad. Sci. U.S.A.* **79**, 203–207.
Webb, C. J. and Bawa, K. S. (1983). Pollen dispersal by hummingbirds and butterflies: a comparative study of two lowland plants. *Evolution* **37**, 1271–1282.
Wright, S. (1948). On the roles of directed and random changes in gene frequency in the genetics of populations. *Evolution* **2**, 279–295.
Wright, S. (1951). The genetical structure of populations. *Ann. Eugen.* **15**, 323–354.

4 Ecological Variation and the Maintenance of Plant Diversity

FAKHRI A. BAZZAZ and SONIA E. SULTAN

Department of Organismic and Evolutionary Biology, Harvard University, Cambridge, MA, U.S.A.

I. Introduction

Experimental studies of plant diversity have focused primarily on genetic variation, estimated by means of electrophoretic samples and DNA restriction-fragment analysis or sequencing. Unfortunately, the interpretation of such variation and its relevance to the organism's success in its environment is not as straightforward as was originally believed (Lewontin, 1974; Ayala, 1982; Schaal, 1985). Although accurate genetic information is a powerful means to discern phylogenetic relationships at a fine scale, genetic variation does not necessarily translate into morphological, physiological, or phenological variation relevant to the fitness of the individual in its environment (Gottlieb, 1981; confirmed experimentally by Giles, 1983).

For this reason we wish to focus attention on ecological variation, here understood to mean the differential performance (i.e. growth, physiology, and reproduction) of a given organism under various environmental conditions. Different plant species, populations, and genotypes may exhibit different growth responses to environmental factors. These characteristic responses have been recognized respectively by ecologists and geneticists as "niches" (*sensu* Bazzaz, 1986), "ecotypes", and "genotype–environment interactions". Ecological variation can be interpreted directly with respect to fitness in local environments, and therefore has clear implications for present distribution and future selective outcomes.

Differentiation Patterns in Higher Plants
ISBN: 0-12-709425-3

Differences between individuals in growth and reproductive success depend not only on nuclear genotype, but also on environmental influences on development (including seed size and quality differences due to maternal plant phenotype and to ovule position during development). For this reason a meaningful description of plant variation in nature must include both plant response to environment and a description of the environment encountered. This is important to evolutionary as well as to ecological questions; variation which may not be entirely genetic in origin will nonetheless have genetic and therefore evolutionary consequences insofar as it affects the number and quality of offspring and the environment of neighbours.

The goal of this paper is to show that experimental studies of ecological variation are an informative and feasible part of research on plant diversity.

II. Patterns of Environmental Heterogeneity

Although general information about environmental conditions in diverse geographical and altitudional regions is available, we know very little about environmental variation in either physical or biotic factors through a successional sequence or at a fine scale within habitats. Data on variability in such factors as soil moisture, nutrient content, light availability, and distribution of mycorrhizae, pollinators, herbivores, seed predators, and pathogens are extremely time-consuming to collect. This is particularly true since sampling must extend over several entire growth seasons to describe adequately patterns of environmental variation encountered by plant populations.

It has long been assumed that open, early-successional habitats are highly variable both spatially and temporally in physical factors such as temperature, light, soil moisture, and nutrients. There have been few attempts to establish this empirically by comparison with other successional stages. We have recently completed a comparison of light and soil moisture variability in three types of community: an early-successional, primarily annual herbaceous community, a late-successional perennial grassland, and a mature deciduous forest (Rice and Bazzaz, unpubl.). The sampling procedure allowed estimation of fine- and broad-scale spatial heterogeneity as well as temporal variability within and between seasons over a three year period. In general, the early-successional fields were more spatially and temporally variable in both light and soil water potential than the grasslands, which in turn were more variable than the forest floor (Fig. 1).

Environmental heterogeneity within habitats is a general property of plant environments, though again it has not been fully characterized empirically. Tilman (1982, Ch. 5) has mapped soil nitrogen content within a 12 m × 12 m plot and found it to vary by 42% within this small area (Fig. 2). Soil nutrient variability occurs within tropical forest gaps caused by tree-falls, since the soil turned up by the exposed roots is much lower in available nitrogen and phosphorous than is the soil beneath either the forest canopy or the fallen trunk, while the soil directly

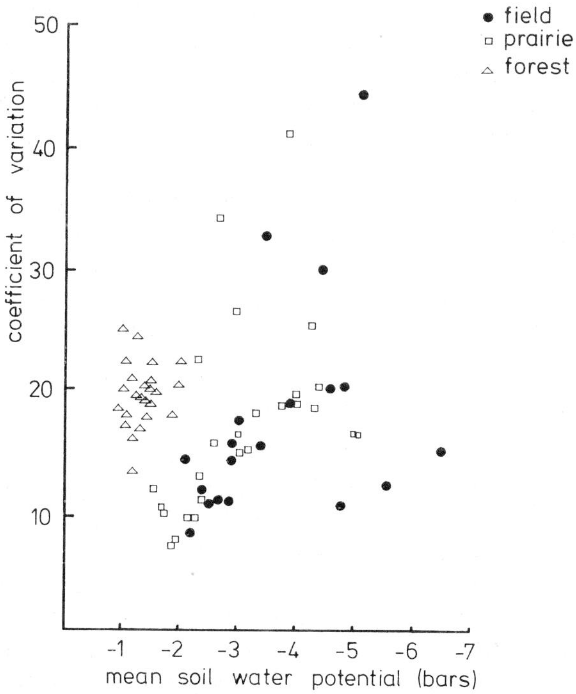

FIG. 1 Variability in soil moisture in field, grassland and deciduous forest over a growing season in mid-western U.S.A. (unpublished data of Rice and Bazzaz).

beneath the crown has enhanced phosphorous and possibly nitrogen content compared with the surrounding forest soil (Bazzaz, 1983; Vitousek and Denslow, unpubl.).

Patchy variation in soil chemistry, microflora, and microfauna can also come about through differences between coexisting species in nutrient uptake and release, accumulation and breakdown of debris, and rhizosphere chemistry, even if the site soil is initially homogeneous (Bazzaz, 1983). Smith (1976) documented significant differences in amino acids, organic acids, carbohydrates, and mineral ions in root exudates from three deciduous forest species. Vegetation also imposes environmental heterogeneity in other factors which influence subsequent plant establishment and stand composition.

Dirzo (1984) has examined the distribution of herbivores as another aspect of spatial patchiness in plant environments. He found very variable distributions of snails and slugs in a pasture within a 50 m × 70 m area sampled (Fig. 3). To the extent that herbivory is either random or correlated with plant characters such as size and apparency that may not be genotypic in origin, differential herbivore damage is an important aspect of environmental heterogeneity in space as well as in time. Similarly, pathogen distribution

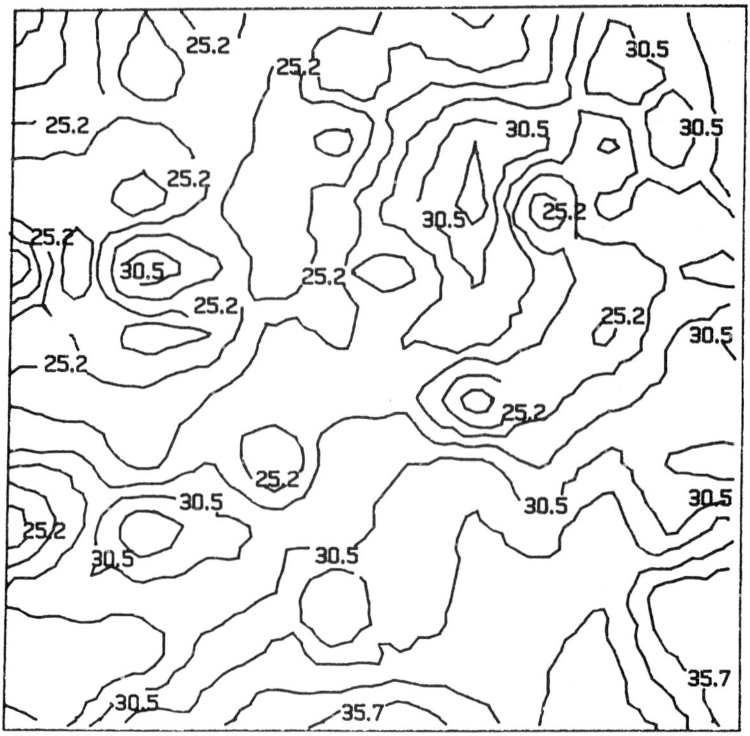

FIG. 2 Spatial heterogeneity in total nitrogen in soils in Minnesota, 12 m × 12 m plot (from Tilman, 1982).

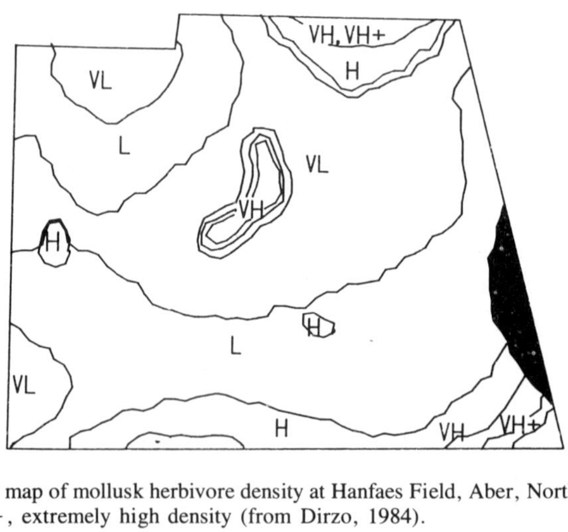

FIG. 3 Contour map of mollusk herbivore density at Hanfaes Field, Aber, North Wales. VL, very low density; VH+, extremely high density (from Dirzo, 1984).

varies according to microsite humidity, and thus comprises a further aspect of environmental patchiness with important fitness implications.

III. Plant Responses on Environmental Gradients

Environmental factors such as light, soil moisture, nutrients, and herbivores vary irregularly on broad to very fine spatial and temporal scales. Different species or populations are located in different large-scale environmental patches and experience different sets of edaphic and ambient conditions. On a finer scale, various sessile individuals occupy different microsites and experience associated differences in temporal variation within and between seasons. Within each individual plant, shoots and roots extend into diverse fine-scale spatial patches, and the individual experiences a shifting set of environmental circumstances through its life span.

Ecological response breadth can be determined experimentally by transforming the natural mosaic of variation in a particular factor into a linear range of controlled environmental states from low to high levels. Such an ordered set of environmental states is termed an environmental gradient. When the naturally occurring environmental patchwork is re-created for experimental purposes as an ordered sequence, plant growth and reproduction in the different states can be characterized as a broad or narrow response curve (Fig. 4). This response curve connotes the actual distribution pattern, since (within the limits of dispersal capability) plants with narrow responses can inhabit only those environmental patches which correspond to a particular point or few points on the gradient, and plants with broad responses may occur in many different types of patch, corresponding to much of the gradient for that factor. Because the environment varies within and between growth seasons as well as spatially within a habitat, actual distribution also reflects the pattern of temporal variability, which elicits particular responses depending on environmental conditions during germination and growth.

A population which as a whole exhibits broad ecological tolerance may consist of a set of diverse, narrowly specialized genotypes or a set of similar, broadly generalized genotypes (Lewontin, 1957; Allard and Bradshaw, 1964). Similarly, a species may consist of specially adapted ecotypic populations or of broadly adapted widespread populations. Information about response patterns can be applied to evolutionary predictions: the extent to which ecological response breadth is divided among separate entities (individuals or populations) strongly influences whether or not selective divergence will occur between these entities.

We have experimentally compared the response patterns of a number of species from different successional communities (e.g., Parrish and Bazzaz, 1979, 1982a,b). In these experiments a sample of individuals from each species is grown on gradients of environmental factors such as light, nutrients, soil moisture, temperature, and density. Their response is measured in terms of germination,

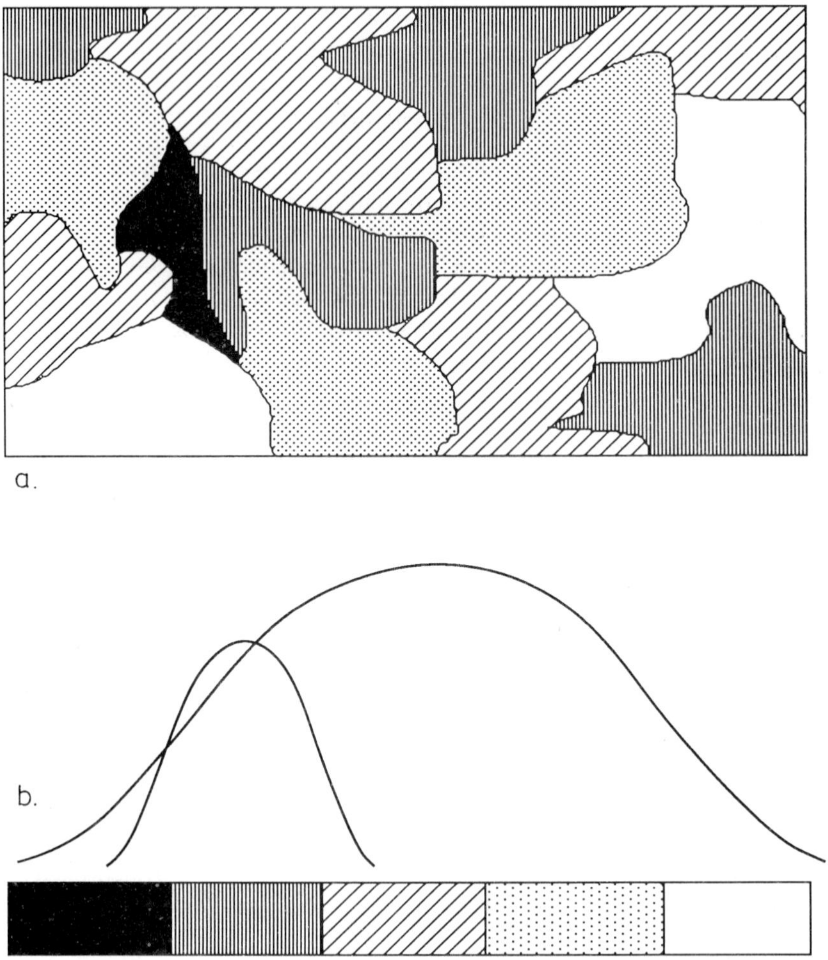

FIG. 4 Representation of patchiness of a resource (a) and rearrangement into a resource gradient (b) allowing comparison of narrow and broad response curves.

survivorship, biomass and reproductive output. Response breadth is characterized for statistical purposes using Levins' *B* formula (Levins, 1968)

$$B = 1/\Sigma(p_i^2)S$$

where *p* is the proportional response in resource state *i* and *S* is the number of resource states. In both the herbaceous and the forest seres studied, the species in the early-successional stages were found to have broader and more overlapping response patterns than those in the late-successional stages (Table 1). When studied, mid-successional species were intermediate. Within these early- and late-successional groups of species, however, each species exhibited a unique response to each environmental gradient. Furthermore, species which showed broad responses to certain factors were not necessarily broad in response to other factors (Table 2). In many species, the range of environments suitable for sexual reproduction is narrower than that in which vegetative growth can occur. In annual species it is clearly the range in which reproduction occurs that is important for fitness.

TABLE 1
Mean response breadth $B = 1/\Sigma(p_i^2)S$ of plants from early- and late-successional communities

	Early succession	Late succession
Herbs		
Underground space	0.71	0.29
Pollinators	0.20	0.16
Nutrients	0.77	0.70
Trees		
Nutrients	0.91	0.81
Moisture	0.89	0.65

TABLE 2
Response breadths (Levins, *B*) of early-successional annuals on four resource gradients

Species	Moisture	Temperature	Light	Nutrients
Abutilon theophrasti	0.88	0.62	0.71	0.60
Amaranthus retroflexus	0.49	0.78	0.83	0.69
Ambrosia artemisiifolia	0.76	0.95	0.91	0.72
Chenopodium album	0.87	0.54	0.88	0.81
Polygonum pensylvanicum	0.99	0.91	0.90	0.82
Setaria faberii	0.99	0.92	0.86	0.76

In certain species the presence and identity of competitors may compress ecological response considerably, while in other species response may be unaffected by competition. For example, *Polygonum pensylvanicum* successfully occupied the full range of soil moisture states when grown without interspecific neighbours, but when grown together with other annual species that occur naturally with it, this species was able to occupy only the very moist state. In contrast, the broad ecological responses of *Ambrosia artemisiifolia* were not at all narrowed in the presence of competing species (Pickett and Bazzaz, 1978). The influence of competition on species response breadth relates directly to observed community diversity patterns. When *A. trifida* seedlings were removed from stands of an annual community including *P. pensylvanicum*, the species diversity of the community increased markedly. The species were able to occupy a large part of their ecological response ranges when not out-competed by the vigorous *Ambrosia* (Abul-Fatih and Bazzaz, 1979).

IV. Genetic Variation in Plant Populations

Much of our thinking regarding both species diversity and population ecology is based on the principle that consistent and strong selection pressure will evoke genetic changes which enable organisms to accommodate environmental demands. This principle has been proven beyond any doubt in studies of strong directional selection for single-gene variants, such as melanism in *Biston betularia* and heavy-metal tolerance in *Agrostis tenuis*, as well as in innumerable artificial selection experiments. Biotic pressures may in certain situations result in selective differentiation within populations, as has been suggested to occur in *Trifolium repens* (Turkington and Harper, 1979). In this case, the results of reciprocal transplant experiments have been interpreted as demonstrating selection for specialized competitive ability.

Given the efficacy of strong, consistent selection as a mechanism of adaptive change, the fundamental question remains as to how strong and consistent selection pressures are in the majority of natural situations (Antonovics, 1976; Sultan, 1987). Under what circumstances will genetic variation for ecologically meaningful characters be winnowed out by natural selection? Under what conditions will directional selection be constrained, and variation maintained? Recent work done in our laboratory and by other workers on ecological variation, environmental heterogeneity, and environmental effects on plant growth and reproductive success, suggests some partial answers to these critical questions.

A brief overview of the evidence regarding genetic variation in natural populations provides the starting point for this discussion. Electrophoretic data compiled over the past two decades indicate remarkably high levels of genetic variation in populations of outcrossing plant species (40–70% polymorphic loci) and similar levels for inbred species when different populations are pooled (Gottlieb, 1981 and references). Relative amounts of genetic variation are largely correlated with breeding system (outcrossed and especially wind-pollinated versus inbred species) and with individual longevity (long-lived perennials and apomicts versus annuals) (Hamrick *et al.*, 1979; Loveless and Hamrick, 1984). In general, then, natural plant populations contain a great deal of genetic variation which is organized largely by population structure.

With respect to selection in such populations, the salient aspect of genetic variation is the existence of variation for characters which directly affect the growth and reproduction of individuals. Despite the strong logical and theoretical expectation that fitness-related characters will have been subject to strong directional selection and therefore will exhibit little genetic variation within populations, there is much evidence that genetic variation for life-history and ecological response characters is abundant. Statistically significant genetic components for life-history variation are typically found in greenhouse and growth chamber studies of natural populations (Venable, 1984).

In many cases in which field experiments are performed in conjunction with experiments under controlled conditions, the genetic components of life-history variation decrease or disappear in the field studies (data compiled in Venable, 1984). It is well known that fitness-related characters generally have low heritabilities in field studies (e.g., Primack and Antonovics, 1981; Solbrig, 1981; Gill *et al.*, 1983; Antonovics, 1984). The environmental homogeneity of greenhouse, common garden, and growth chamber studies reveals the existence of genetic variability for fitness-related characters; however genotypic differences appear to be largely masked by phenotypic responses to environmental variation in the field. Selection may thus be impeded from removing particular life-history variants by the confounding and often overwhelming effects of environmental heterogeneity.

V. Environmental Variation and Diversity

Environmental heterogeneity is thought to promote species diversity in plant communities (Whittaker, 1975). Temporal patterns in species diversity have been described as a repeating wave which is re-initiated by random perturbations, such as forest fires, at intervals of several to several hundred years (Loucks, 1970). Through a successional sequence which increases in floristic diversity from old-field annuals to deciduous forest, increasing structural complexity generates vertical and horizontal microenvironmental variation which can be exploited by species of different ecological response patterns (Bazzaz, 1975). Grubb (1977) and others have argued that species richness in plant communities may be enhanced by environmental patchiness generated by disturbance. Events such as limb breakage or tree falls create gap microhabitats which differ in illumination, nutrient availability, depth of soil upheaval, etc. Depending on seed distribution and plant growth requirements, different species may colonize these gap microhabitats. Fluctuation in rainfall and temperature will elicit different dominant species in different years, so that more than one dominant species is maintained in late-successional grasslands (Grubb, 1977).

At the population level, too, environmental patchiness contributes to the maintenance of diversity. A great deal of theoretical work has explored the role of environmental variability in maintaining genetic variation (see Hedrick *et al.*, 1976; Ennos, 1983; Orzack, 1985; Rice and Jain, 1985, for references). Although it is not our purpose here to review this extensive literature in detail, we very briefly summarize several relevant results and consider their applicability to plant populations.

Several models have shown that (except under extremely restrictive conditions such as regular cycles) temporal variation in selective pressures hastens rather than prevents the loss of variant alleles (Hedrick *et al.*, 1976 and references). However, since the vast majority of plant species are long-lived and iteroparous and/or produce long-lived (dormant) propagules, genetic longevity is usually much greater than the scale of environmental variation. Furthermore, environmental fluctuations generally do not result in differential mortality among genotypes, but in differential reproductive success in particular seasons. For these reasons, temporal variation within and between seasons acts primarily to interfere with the fitness constancy of individual genotypes, rather than to produce a drift-like loss of variants.

Theoretical as well as empirical studies have shown that selection can adaptively subdivide populations which inhabit spatially diverse areas, if the spatial variation is constant (coarse-grained), selective differentials are large, and gene flow is restricted (summary and references in Ennos, 1983). This type of selection for multiple niches depends on constant fitness differences, and has been demonstrated unequivocally only in a case where consistent environmental pressures were imposed experimentally (Snaydon and Davies, 1972).

The population genetics scenario which we consider most applicable to plant populations in natural environments is that of frequency dependence. When fitnesses are not fixed but contingent on relative competitive success (as determined by the individual's genotype and microsite, and the genetic and microenvironmental composition of the neighbourhood), complex and generally unstable polymorphisms will be maintained (Ennos, 1983 and references). Frequency-dependent selection favouring rare genotypes is also imposed by herbivores and pathogens which seek common morphs (e.g., Cahn and Harper, 1976) or have evolved immunity to existing plant defences (Burdon, 1985). The spatial and temporal variability that characterizes environments occupied by plant populations imposes selective pressures which are complex and fluctuating rather than one-directional and consistent. This aspect of the plant environment imposes concomitant variation in relative fitness which promotes the maintenance of genetic variation; furthermore, it increases the fitness value of genotypes which confer broad ecological responses.

This issue was addressed in a recent study of *Amaranthus retroflexus*. Individuals of this selfing species are homozygous for one of two electrophoretic alleles at the phosphoglucoisomerase (PGI) locus. The two genotypes entail markedly different patterns of response to soil moisture, such that one is superior at high soil moisture and the other at medium soil moisture (Zangerl and Bazzaz, 1984b; Fig. 5). Although the relationship between genetic variation and individual fitness differences under particular conditions is well established in this case, and the relevant genetic variation does exist in the populations examined, both genotypes are maintained in the populations. Variation within sites and between years in soil moisture content ranges from medium to high levels, and therefore (since seeds remain alive in the soil for a number of years) neither variant is eliminated by natural selection.

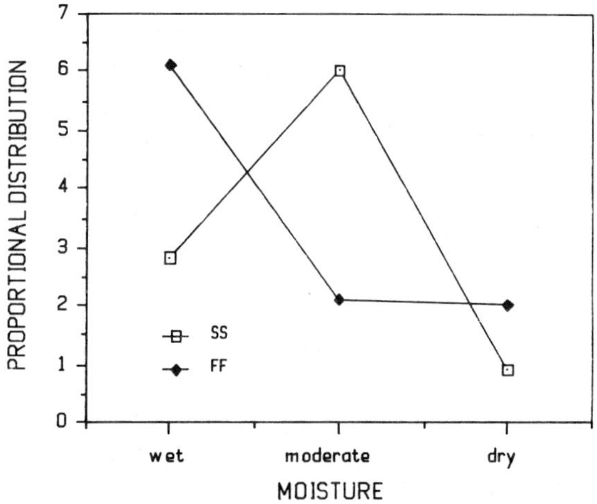

FIG. 5 Proportional distribution of total seed biomass in each of the two PGI homozygous genotypes grown together along the soil moisture gradient. The sum of proportions over all three states equals 1.0 for each genotype.

Fitness differences under controlled conditions were demonstrated between white and pigmented individuals of wild *Phlox drummondii* populations. Although the colour morphs differ at only a few loci, white plants are less vigorous and have high mortality rates in dense stands, as compared with pigmented individuals. Despite this fitness differential, white morphs comprise 4% of the natural populations, which are known to be at least 50 years old (Bazzaz *et al.*, 1982). Persistence of demonstrably less fit genotypes has also been shown to occur in *Solanum dulcamara* (Clough *et al.*, 1980).

A second study of *P. drummondii* (Schwaegerle *et al.*, 1986) examined between-population genetic variation and response patterns in nine populations from contrasting sites. These sites differed in temperature regime, soil moisture, nutrients, and vegetation density — environmental factors which vary and interact within sites. Although the populations were found to be genetically distinct (as measured by both electrophoresis and quantitative genetics), the different ecological response patterns which characterized the populations did not fully correspond to the environmental differences between the population sites: plants from nutrient-poor sites grew poorly in low-nutrient conditions as compared with plants from other populations, plants from soils with low water-holding capacity were less drought-tolerant than those from other sites, etc. (Schwaegerle and Bazzaz, 1987). Although in this case the constraints on selection may be partly genetic (since the strong between-population divergence may reflect genetic drift), these data suggest that the action of moderate and complex selection pressures may fail to produce complete adaptation to local environment.

A neglected aspect of environmental variation is that of community composition and stand density. If consistent, these biotic factors may exert strong selective pressures on plant populations for competitive ability, shade tolerance, and phenological displacement. However, change between seasons in community composition may buffer such selection. Greenhouse studies of a mid-western U.S.A. annual community showed that a given mixture of sown seeds produced different proportions of the component species when grown at different moisture and nutrient levels (Pickett and Bazzaz, 1978; Parrish and Bazzaz, 1982a, 1985b). When different spring conditions were simulated to mimic naturally occurring between-season variation in temperature, soil moisture, soil topography, and soil nutrients, identically sown "seed banks" (representing the mid-western U.S.A. and a British annual community) produced dramatically different proportions of component species in terms of per cent biomass (Bazzaz and Grubb, unpubl.; Fig. 6). Because different annual species have different germination, growth, and fecundity responses to different environmental conditions, variation in conditions during early spring within sites and between seasons will result in patches which are dominated by different species at different times, and which leave a record of this fluctuation in the seed bank. Variation in abundance of annuals between years is well documented: for example, Grubb *et al.* (1982) censussed stands of *Linum catharticum* on a chalk grassland in Sussex for three years and found them to vary in number from 78 to 7700 individuals in one site and from 102 to about 15 000 in another.

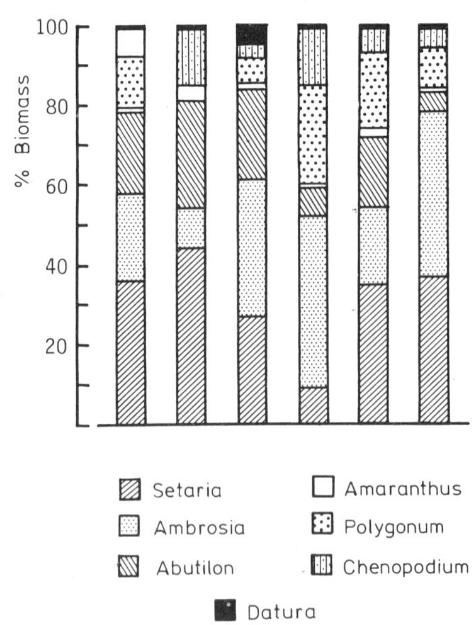

FIG. 6 Proportional contribution of various species to total community biomass under six growth conditions (from left to right): control, drought, shade, cool, rough surface, and unfertilized treatments.

When artificial selection experiments were conducted on *Amaranthus retroflexus* using different levels of community diversity as selective regimes, this effect of community variability was inadvertently produced: generations grown in the greenhouse at different times of the year (i.e. with different temperatures) encountered different proportions of the neighbour species which comprised the treatments. Although lines selected on high and low soil moisture treatments each became fixed for a single PGI allele, both alleles were maintained after four generations in the diversity experiment, suggesting that the lines were not in fact under constant environmental pressures (Zangerl and Bazzaz, 1984a).

In annual species with dormant seeds, temporal variation between seasons may hinder selection for specialization in germination response to particular conditions. In the winter annual *Collinsia verna*, rosette size at the onset of winter as well as adult fecundity differences are strongly correlated with germination (emergence?) timing. Despite the influence of germination timing on plant fitness, both the magnitude and the direction of selection for germination timing were shown to vary between the two seasons studied, as determined by regression analysis of phenotype distribution (Kalisz, 1986). Although genotypes which perform particularly well under given conditions will produce a disproportionate amount of the seed contribution for that season (Solbrig and Solbrig, 1984), it can be shown mathematically (Brown and Venable, 1986) that if many different kinds of season occur unpredictably and in varying frequency, evolutionary specialization for seedling response will be opposed. In addition, an increase in per cent germination in any year — as happens for example with occasional disturbance — increases the variance in yearly fitness and thereby opposes selection for specialization (Brown and Venable, 1986).

VI. Other Ecological Mechanisms Important to the Maintenance of Genetic Variation

Several general properties of plants as individuals and as populations promote the maintenance of genetic variation in the context of heterogeneous environments. These are discussed briefly below. (For a summary of genetic mechanisms in constant environments, such as heterosis, meiotic drive, and incompatibility systems, see Ennos (1983).)

A. Individual response flexibility

Genetic individuals in many plant species exhibit much flexibility in size, morphology, and physiological behaviour in response to environmental variation. This phenotypic plasticity contributes to the tremendous difficulties encountered by plant taxonomists at the specific and subspecific levels. In many species, plastic responses appear to promote plant fitness under the conditions that elicit them, and may in fact represent selected response capabilities (Bradshaw, 1965; Jain, 1979; Schlichting, 1986). In these cases, the flexible expression of individual genotypes comprises a mechanism whereby the population can accommodate environmental stresses without selection for genetic subdivision or directional change. In this way existing genetic variation may be maintained.

Within-genotype plasticity is well known in root, shoot, and reproductive morphology, size, anatomy, relative and absolute biomass allocation, and photosynthetic rates, as well as in facultative growth responses such as many cases of dormancy, deciduous habit, shoot dimorphism, and cleistogamy. Here we will cite several examples of plasticity which appears to mitigate environmental pressure and thus to buffer selection (see Sultan, 1987 and references for full discussion).

In densely crowded plant populations, mortality (and genetic deaths) may be avoided if individuals reduce growth rates and thus "absorb" density stress through reduction in size (Harper, 1977). If size reduction in annual plants is accompanied by a proportional increase in allocation to reproduction relative to biomass, reproductive output and fitness is maintained at levels similar to those of uncrowded plants. In *Ambrosia trifida*, lack of branching, low leaf turnover, and proportionally greater biomass allocation to reproductive structures in densely crowded individuals result in low mortality and maintenance of high seed production (Bazzaz, 1984). A similar result was found in *Polygonum cascadense*, in which plants in dense stands showed a 70% increase in reproduction relative to biomass as well as lower seedling mortality compared to those in stands of rapidly growing uncrowded individuals (Hickman, 1975).

In *Ambrosia trifida*, response to density involves an unusual kind of developmental flexibility. In dense stands, the suppressed (later-emerging) individuals of this normally monoecious species produce only female flowers. Since functionally female plants are surrounded by pollen-producing neighbours, this plasticity in sex expression permits suppressed individuals to invest proportionally more energy into seed production (although they also lose the male function component of fitness). The proportion of total biomass allocated to seed in these functionally female plants is 10% as compared with 3% in the larger, bisexual individuals (Abul-Fatih *et al.*, 1979).

Increased relative biomass allocation to roots is another means by which plant flexibility can mediate environmental stress. In several species, relative allocation to roots increases under conditions of low soil moisture. Changes in leaf angle and petiole length can permit certain plants to vary leaf temperature with respect to unfavourable ambient conditions. *Abutilon theophrasti* is able to maintain leaf temperatures close to its photosynthetic optimum when air temperature is excessively high by varying leaf angles up to 60° above horizontal. This reduces leaf temperature by 8°C when the sun is overhead (Bazzaz, 1979). Similar orientation behaviour has been noted in the desert plant *Atriplex hymenelytra* (Mooney and Gulmon, 1979).

Physiological as well as morphological plasticity allows individuals in many species to accommodate both sun and shade. Such species produce thick "sun leaves" which exhibit high photosynthetic rates in well-lit situations, and thin "shade leaves" which have low light-compensation points and concomitant low photosynthetic rates in shaded microsites (see references in Sultan, 1987). Some species are able to switch to CAM metabolism under conditions of drought stress, and to maintain the more efficient C_3 pathway if water is not limiting (Mooney and Gulmon, 1979).

Defence capabilities are also plastic in some species that synthesize defence compounds facultatively if subjected to pathogen or herbivore attack (Ryan, 1979, 1983; McKey, 1979; Crawley, 1983; Rhoades, 1983 and references therein). Whitham (1983) has focused attention on cases of pronounced quantitative and/or qualitative variability in chemical defence within long-lived plant individuals. Such variation may result from induced responses to unevenly distributed herbivory, ontogenetic variation in plant parts of different age, or to genetic mosaicism due to somatic mutation. This within-plant variability in defence may prevent herbivores from breaking the defences of the plants, and thus obviate further defensive evolution by the host species. Although the hypothesis of within-plant defensive evolution by somatic mutation implies a different evolutionary outcome in the plant population, the mechanisms of observed variation within plant individuals are not known (Whitham, 1983), and there is no direct evidence of genetic variation within plants for chemical defence properties.

B. Clonal integration

While unitary plant growth forms can mitigate fine-scale environmental heterogeneity by developmental and physiological plasticity, genetic individuals in many clonal species are able to integrate spatial environmental variation encountered by different ramets. Physiological integration among ramets of a clone has been found in a number of species in which photosynthate is translocated from the parental clone to developing ramets, thus reducing the probability of ramet mortality in unfavourable microsites (Anderson-Taylor and Marshall, 1983; Hartnett and Bazzaz, 1983; Ashmun *et al.*, 1982 and references). In a study of *Solidago canadensis*, shaded ramets exhibited 100% survivorship if connections to illuminated sibling ramets were maintained, but fell quickly to 80% if the shaded ramets had been severed from sibling ramets (Hartnett and Bazzaz, 1983). This difference was apparent only in the first seven weeks of the season, indicating that early in the season (before *Solidago* will have overtopped its neighbours) this kind of clonal integration affords a means to avoid the consequences of ramet shading. Similar results have been obtained in shading and defoliation experiments on grass species (Marshall and Sagar, 1968). When dense populations of interconnected ramets were compared with dense populations of independent *Solidago* seedlings, the reduction in biomass with increasing density was much less steep in the interconnected ramets (slope of -0.63 versus -0.98 on a log biomass versus log density plot) (Hartnett and Bazzaz, 1985c). Connected ramets were also more equitable in size than independent plants grown at low, medium, and high densities, indicating that the integrated ramet system was less susceptible to growth constraints due to high density.

Clonal integration among ramets has been shown experimentally to buffer variation in biotic neighbourhood as well. Shoots of different clonal genets of *Solidago canadensis* had significantly different biomass and reproductive responses if the genets were grown with only one neighbour species. When each genet had several neighbour species — as in a natural clone in which ramets might encounter different neighbours — its ramets had very similar growth and reproductive responses (Hartnett and Bazzaz, 1985b). When genets of *Bellis perennis* were grown on a density gradient so that sets of ramets encountered different neighbour densities, the genet as a whole apparently integrated this environmental patchiness such that the clone expanded equally in all directions. In contrast, the loose, meandering growth form of *Prunella vulgaris* clones responded to the density gradient by enhanced growth of those ramets encountering sparse neighbours and reduced growth of ramets in dense neighbourhoods (Schmid, 1986). Thus clonal species may vary in the extent to which they integrate environmental heterogeneity, depending on physiology and growth form. The evolution of individual response flexibility should be favoured in clonal species with a tightly packed or "phalanx" growth form relative to "guerilla" species which have wandering ramets that can seek out favourable patches. Greater plasticity and ecological tolerance of the former has been found in several comparisons of these two kinds of clonal species (Schmid, 1985 and references).

As genets cover increasingly large areas with age, they are likely to encompass more of the local environmental variation in patch types. Physiological integration of ramets should buffer genets against patch-specific fine-scale selection pressures, and thus distribute the probability of success evenly among different genotypes (Hartnett and Bazzaz, 1985a). This should result in maintenance of most of the genotypes that originally colonized a site. This prediction was tested in further studies of *Solidago canadensis*. Genetic diversity was apparently maintained in the field: of twenty clones monitored for a five year period, no genet mortality occurred. These twenty genets appeared morphologically and phenologically different, and as this is an obligately outbred species, were presumably genetically distinct (Hartnett and Bazzaz, 1985a). Persistence of genetic diversity has also been observed in *Athyrium filix-femina* (Schneller and Schmid, 1982). In species with less physiological integration, population genetic diversity may be lower (but see Harper, 1985 for an opposing view).

Perennial plants which do not have a clonally spreading growth form integrate environmental variability in time rather than in space. Because perennials have storage organs and, in woody species, cumulative growth, and because reproductive allocation can vary from favourable to unfavourable seasons, perennials are buffered from both within- as well as between-season environmental variability.

C. Environmentally imposed determinants of fitness

Because plant growth and reproductive output vary dramatically within populations due to environmental circumstances, and because plants as sessile organisms cannot seek appropriate microsites, variation in fitness does not necessarily reflect genotypic differences. Non-heritable influences on plant fitness may obscure or override genotypic differences and thus oppose directional or disruptive selection (Sultan, 1987).

Under nutrient, light, or drought stress, an individual plant may produce fewer and/or smaller seeds (Harper, 1977). Certain species may reduce seed size rather than number, while in other species size may be quite constant. The effect of seed size variation on probability of emergence, seedling survival and growth rate, and adult plant size and reproductive output is well known (Stanton, 1984; Schaal, 1984; Antonovics and Schmitt, 1986 and references in these papers). Schaal (1984) has reviewed evidence that such variation in seed size may be due to the environmental circumstances of the maternal plant. Such effects may carry over several generations. The temperature of the maternal environment affects seed size, per cent germination, and seedling size in several species (see references in Alexander and Wulff, 1985). In *Plantago lanceolata*, such maternal effects were detected in the second generation after treatment. Since these plants flower over a 2–3 month period in their native habitat in North Carolina, differences in temperature during ovule development are naturally encountered (Alexander and Wulff, 1985). Parrish and Bazzaz (1985a) demonstrated that in *Abutilon theophrasti* different maternal nutrient conditions affected not only seed size, but also seed nutrient content, and led to different competitive abilities among seedlings.

Since seed size reflects paternal and maternal genetic contributions, non-additive properties of the zygotic genotype, maternal and grand-maternal environment, and position of the developing ovule on the maternal plant, correct attribution of seed size variation requires meticulous experimentation. In a recent study of *Anthoxanthum odoratum*, Antonovics and Schmitt (1986) found that 80% of total variation in propagule size was attributable to environmental effects, largely within maternal plants (i.e. due to different ovule positions). Field data confirmed these results. There may indeed be a great deal of seed size and progeny fitness variation within maternal plants due to ovule position effects, which is underestimated due to inadequate within-individual sampling. Seed size variation is also underestimated because seeds are usually weighed in groups to determine mean size (Stanton, 1984).

The favourable or unfavourable microsite to which a seed is dispersed has a critical effect on germination probability, seedling survivorship and growth rate, and adult plant size and fecundity (Harper, 1977). Differences at a minute (single-seed) scale in soil compaction, water and nutrient availability, and presence or absence of physical obstructions below or at the soil surface, have been shown to strongly influence seed germination and subsequent plant growth (Harper *et al.*, 1965; Hartgerink and Bazzaz, 1984). Microsite variation at this scale can be introduced by animal activity (e.g. footprints, burrows, droppings), by falling plant debris, etc., and is clearly both randomly distributed and unpredictable.

In addition, the depth of seed burial and the pattern of soil disturbance largely determines when a seedling will emerge and how much of a nutrient reserve it retains upon emergence to expend on growth. Emergence timing is an important determinant of competitive success and adult fitness, since late emergents are often at a pronounced disadvantage with respect to survival and ultimately fecundity. In the winter annual *Geranium carolinianum*, seeds produced on the same plant over a four-week period varied in size and germination timing: later-produced seeds were nearly 50% smaller, and germinated more quickly. These earlier germinants had greater rosette development before the onset of winter and were larger and more fecund as adults the following spring (Roach, 1986). In the winter annuals *Erigeron canadensis* and *E. annuus*, the competitive superiority of *E. annuus* is largely attributable to its generally earlier seedling emergence. Earlier emergents produce larger rosettes which have a much higher probability of surviving over winter, more rapid growth rates in early spring, and larger contributions to the seed pool (Regehr and Bazzaz, 1979; Bazzaz, 1984).

Depending on random seed dispersal and burial, habitat patchiness, and disturbance regimes, populations will contain different age classes of individuals. Ontogenetic shifts in ecological response create within-population variation between these individuals of different ages that may obscure genotype-specific differences which might otherwise lead to selective elimination of genotypes. Plants at different ontogenetic stages may have different responses to nutrient states, temperatures, and drought stress (Parrish and Bazzaz, 1985b and references) and competitive abilities with respect to co-occurring species (Grace, 1985). Ecological responses of perennials may change as they develop from seedlings to rhizomatous shoot systems. In *Polygonum virginianum*, seedlings and plantlets from rhizome cuttings have quite distinct response patterns. Rhizome-derived plants are more successful than seedlings in nutrient-poor treatments (Zangerl and Bazzaz, 1983). Responses in terms of leaf shape and thickness, internode and petiole length, and other morphometric characters were significantly different in seedlings and rhizome-derived plants on moisture and light gradients as well (Lee *et al.*, 1986).

Patterns of herbivory are another important determinant of differential fitness which may not correspond to genetic differences, but which will have significant effects on the genetic composition of subsequent generations. Although in some systems herbivory and seed predation are correlated with genetic variation for defensive chemistry, in other cases herbivore distribution and behaviour, and plant location and apparency, will produce patterns of differential predation that are not genetically correlated. Reduction of fitness may be direct, due to loss of meristems, flowers, fruits or seeds, or indirect due to loss of vegetative biomass which subsequently reduces reproduction (Dirzo, 1984). Facultative biosynthetic responses to herbivory may also entail indirect costs to a plant under predation. Simulated herbivory in the form of seed damage demonstrated dramatically the impact of herbivory on fitness: as little as 5% seed damage resulted in notably reduced seedling growth rates (Dirzo, 1984).

It has been argued that when a better competitor is also a preferred food species, herbivory can function to maintain community diversity by offsetting the competitive differential (Paine, 1966; Harper, 1969; Caswell, 1978 and references; Dirzo, 1984). This can be true at the population level as well: if a more vigorous genotype produces larger seeds, more attractive foliage, and higher plant apparency to herbivores, it may suffer increased herbivory and a resulting reduction in fitness. Herbivory may thus maintain genetic diversity for components of fitness that are positively correlated genetically, developmentally, or ecologically with increased predation, by offsetting other selective advantages or augmenting other random aspects of environmental variation.

D. *Life-history characteristics*

Several aspects of plant life-history promote the maintenance of genetic variation in populations. This is not necessarily an evolutionary "strategy": in fact selection for adaptive response to changes in environment may be opposed by this kind of temporal gene flow.

Apomixis protects entire genomes, and in mixed systems maintains allelic diversity for recombination. Apomixis thus increases effective population size by storing genetic variation (Levin and Kerster, 1971). Long-lived clones and iteroparous perennials similarly hold genetic variation in populations over many seasons while environmental conditions may fluctuate or change directionally.

The existence of a seed bank for the many annual species with dormant seeds signifies that populations in the sense of genetic pools are much larger and more diverse than they appear at any one time. This storage of genetic variation prevents genetic losses due to drift (Loveless and Hamrick, 1984). In addition, fluctuations in environment which may periodically reduce the size of standing populations do not result in genetic bottlenecks, as variation is stored in the underground seed population. This allows the population to resist genetic response to short-term environmental fluctuation (Templeton and Levin, 1979). However, the seed bank also constrains evolutionary response to consistently changing conditions (Templeton and Levin, 1979). For example, genetic subdivision in *Veronica peregrina* populations failed to occur in sites with soil disturbance every spring, although it did occur very rapidly in similar sites without disturbance and consequent mixing of seeds (Linhart, 1974).

Immigration into populations by means of pollen and seed dispersal is an important means of maintaining genetic variation (see Levin, 1984 for a recent review of this issue). Variants previously lost, either by founder effects and drift or by local selection, can be re-introduced from nearby populations. Continued gene flow was clearly demonstrated in the case of re-colonization of contaminated sites by metal-intolerant genotypes of *Agrostis tenuis*. This did not effectively dilute the tolerant population that had arisen, because in this case the selection pressure was 100% effective against the immigrants (McNeilly and Antonovics, 1968). However, pollen and seed flow are very effective introducers of genetic variation when selection coefficients are moderate (Antonovics, 1968).

Unfortunately, gene flow into natural populations is exceptionally difficult to measure and therefore very poorly understood (Levin, 1984). Estimates based on pollinator flight distances appear to be inadequate, since pollen carry-over may occur over relatively large distances in pollinators with directional flight paths (Levin, 1981). Although pollen and seed dispersal are largely restricted to areas adjacent to parental plants (Ehrlich and Raven, 1969; Levin and Kerster, 1974), the tail end of this skewed dispersal curve is never ascertained. Since genetic divergence of populations can be prevented by very few immigrants in each generation (Wright, 1931), the importance of these few long-distance dispersal events for the maintenance of genetic variation should not be underestimated.

VII. Conclusions

Ecological variation is a direct means of evaluating the performance of plants in their environments. The study of plant response across controlled environmental axes affords valuable insight into plant distribution and diversity at several levels.

There is no question that natural selection plays a central role in many cases of within- and between-population divergence and therefore, ultimately, in species formation. However, the complexity of interactions in nature between the heterogeneous plant environment, the plant's response pattern, and the structuring influence of gene flow, may mitigate the force of selection. In order to refine our understanding of selection and thus diversity, we must consider the many factors that interact with and mitigate selection in nature. In this chapter we have identified factors which act to maintain genetic and species diversity: environmental variation and its effects on fitness, individual plasticity, clonal integration, and certain life-history characteristics. A great deal of important work remains to be done to explore spatial and temporal patterns of environmental heterogeneity in biotic and physical factors, the immediate and evolutionary impact of this heterogeneity on plants, flexible means of individual response, sampling of the environment by clonally spreading species, and the extent of gene flow within and between populations.

References

Abul-Fatih, H. A. and Bazzaz, F. A. (1979). The biology of *Ambrosia trifida* L. I. Influence of species removal on the organization of the plant community. *New Phytol.* **83**, 813–816.

Abul-Fatih, H. A., Bazzaz, F. A. and Hunt, R. (1979). The biology of *Ambrosia trifida* L. III. Growth and biomass allocation. *New Phytol.* 83, 829–838.

Alexander, H. M. and Wulff, R. D. (1985). Experimental ecological genetics in *Plantago*. X. The effects of maternal temperature on seed and seedling characters in *P. lanceolata*. *J. Ecol.* **73**, 271–282.

Allard, R. W. and Bradshaw, A. D. (1964). The implications of genotype–environment interactions in applied plant breeding. *Crop Sci.* **4**, 503–508.

Anderson-Taylor, G. and Marshall, C. (1983). Root–tiller inter-relationships in spring barley (*Hordeum distichum* (L.). Lam.). *Ann. Bot.* **51**, 47–58.

Antonovics, J. (1968). Evolution in closely adjacent plant populations. VI. Manifold effects of gene flow. *Heredity* **23**, 507–524.

Antonovics, J. (1976). The nature of limits to natural selection. *Ann. Mo. Bot. Gard.* **63**, 221–247.

Antonovics, J. (1984). Genetic variation within populations. *In* "Perspectives in Plant Population Ecology" (R. Dirzo and J. Sarukhan, eds), pp. 229–241. Sinauer Associates Inc., Sunderland. MA.

Antonovics, J. and Schmitt, J. (1986). Paternal and maternal effects on propagule size in *Anthoxanthum odoratum*. *Oecologia* **69**, 277–282.

Ashmun, J. W., Thomas, R. J. and Pitelka, L. F. (1982). Translocation of photoassimilates between sister ramets in two rhizomatous forest herbs. *Ann. Bot.* **49**, 403–415.

Ayala, F. (1982). The genetic structure of species. *In* "Perspectives in Evolution" (R. Milkman, ed.), pp. 60–82. Sinauer Associates Inc., Sunderland, MA.

Bazzaz, F. A. (1975). Plant species diversity in old-field successional ecosystems in southern Illinois. *Ecology* **56**, 485–488.

Bazzaz, F. A. (1979). Physiological ecology of plant succession. *Ann. Rev. Ecol. Syst.* **10**, 351–371.

Bazzaz, F. A. (1983). Characteristics of populations in relation to disturbance in natural and man-modified ecosystems. *In* "Disturbance and Ecosystems — Components of Response" (H. A. Mooney and M. Godron, eds), pp. 259–275. Springer, Berlin.

Bazzaz, F. A. (1984). Demographic consequences of plant physiological traits: some case studies. *In* "Perspectives in Plant Population Ecology" (R. Dirzo and J. Sarukhan, eds), pp. 324–346. Sinauer Associates Inc., Sunderland, MA.

Bazzaz, F. A. (1987). Experimental studies on the evolution of niche in successional plant populations: a synthesis. *In* "Colonization, Succession and Stability" (A. Gray, N. Crawley and P. J. Edwards, eds), pp. 245–272. Special Publication Series of the British Ecology Society, 26. Blackwell Scientific, Oxford.

Bazzaz, F. A., Levin, D. A. and Shmierbach, M. R. (1982). Differential survival of genetic variants in crowded populations of *Phlox*. *J. Appl. Ecol.* **19**, 891–900.

Bradshaw, A. D. (1965). Evolutionary significance of phenotypic plasticity in plants. *Adv. Genet.* **13**, 115–155.

Brown, J. S. and Venable, D. L. (1986). Evolutionary ecology of seed-bank annuals in temporally varying environments. *Am. Nat.* **127**, 31–47.

Burdon, J. J. (1985). Pathogens and the genetic structure of plant populations. *In* "Studies on Plant Demography: A Festschrift for John Harper" (J. White, ed.), pp. 313–325. Academic Press, New York.

Cahn, M. G. and Harper, J. L. (1976). The biology of the leaf mark polymorphism in *Trifolium repens* L. 2. Evidence for the selection of leaf mark by rumen fistulated sheep. *Heredity* **37**, 327–333.

Caswell, H. (1978). A general formula for the sensitivity of population growth rate to changes in life history parameters. *Theor. Popul. Biol.* **14**, 215–230.

Clough, J. M., Teeri, J. A. and Alberte, R. S. (1980). Photosynthetic adaptation of *Solanum dulcamara* L. to sun and shade environments. III. Characterization of genotypes with differing photosynthetic performance. *Oecologia* **44**, 221–225.

Crawley, M. J. (1983). "Herbivory: The Dynamics of Animal–Plant Interactions". University of California Press, Berkeley, CA.

Dirzo, R. (1984). Herbivory: a phytocentric overview. *In* "Perspectives in Plant Population Ecology" (R. Dirzo and J. Sarukhan, eds), pp. 141–165. Sinauer Associates Inc., Sunderland, MA.

Ehrlich, P. R. and Raven, P. H. (1969). Differentiation of populations. *Science* **165**, 1228–1232.

Ennos, R. A. (1983). Maintenance of genetic variation in plant populations. *Evol. Biol.* **16**, 129–155.

Giles, B. E. (1983). A comparison between quantitative and biochemical variation in the wild barley *Hordeum murinum. Evolution* **38**, 34–41.

Gill, D. E., Berven, K. A. and Mock, B. A. (1983). The environmental component of evolutionary biology, *In* "Population Biology: Retrospect and Prospect" (C. E. King and P. S. Dawson, eds), pp. 1–36. Columbia University Press, New York.

Gottlieb, L. D. (1981). Electrophoretic evidence and plant populations. *Progr. Phytochem.* **7**, 1–46.

Grubb, P. J. (1977). The maintenance of species richness in plant communities: the importance of the regeneration niche. *Biol. Rev.* **52**, 108–145.

Grubb, P. J., Kelly, D. and Mitchley, J. (1982). The control of relative abundance in communities of herbaceous plants. *In* "The Plant Community as a Working Mechanism" (E. I. Newman, ed.), pp. 79–98. Special Publication Series of the British Ecology Society, 1. Blackwell Scientific, Oxford.

Grace, J. B. (1985). Juvenile vs. adult competitive abilities in plants: size-dependence in cattails (*Typha*). *Ecology* **66**, 1630–1638.

Hamrick, J. L., Linhart, Y. B. and Mitton, J. B. (1979). Relationship between life-history characteristics and electrophoretically detectable genetic variation in plants. *Ann. Rev. Ecol. Syst.* **10**, 173–200.

Harper, J. L. (1969). The role of predation in vegetational diversity. Diversity and stability in ecological systems. *Brookhaven Symp. Biol.* **22**, 48–62.

Harper, J. L. (1977). "Population Biology of Plants". Academic Press, London.

Harper, J. L. (1985). Modules, branches, and the capture of resources. *In* "Population Biology and Evolution of Clonal Organisms" (J. B. C. Jackson, L. W. Buss and R. E. Cook, eds), pp. 1–33. Yale University Press, New Haven, CT.

Harper, J. L., Williams, J. T. and Sagar, G. R. (1965). The behaviour of seeds in soil. 1. The heterogeneity of soil surfaces and its role in determining the establishment of plants from seed. *J. Ecol.* **53**, 273–286.

Hartgerink, A. P. and Bazzaz, F. A. (1984). Seedling-scale environmental heterogeneity influences individual fitness and population structure. *Ecology* **65**, 198–206.

Hartnett, D. C. and Bazzaz, F. A. (1983). Physiological integration among intraclonal ramets of *Solidago canadensis. Ecology* **64**, 779–788.

Hartnett, D. C. and Bazzaz, F. A. (1985a). The genet and ramet population dynamics of *Solidago canadensis* in an abandoned field. *J. Ecol.* **73**, 407–413.

Hartnett, D. C. and Bazzaz, F. A. (1985b). The integration of neighborhood effects by clonal genets of *Solidato canadensis. J. Ecol.* **73**, 415–427.

Hartnett, D. C. and Bazzaz, F. A. (1985c). The regulation of leaf, ramet and genet densities in experimental populations of the rhizomatous perennial *Solidago canadensis. J. Ecol.* **73**, 429–443.

Hedrick, P. W., Ginevan, M. E. and Ewing, E. P. (1976). Genetic polymorphism in heterogeneous environments. *Ann. Rev. Ecol. Syst.* **7**, 1–32.

Hickman, J. C. (1975). Environmental unpredictability and plastic energy allocation strategies in the annual *Polygonum cascadense. J. Ecol.* **63**, 689–701.

Jain, S. (1979). Adaptive strategies: polymorphism, plasticity, and homeostasis. *In* "Topics in Plant Population Biology" (O. T. Solbrig, S. Jain, G. B. Johnson and P. H. Raven, eds), pp. 160–187. Columbia University Press, London and Basingstoke.

Kalisz, S. (1986). Variable selection on the timing of germination in *Collinsia verna* (Scrophulariaceae). *Evolution* **40**, 479–491.

Lee, H. S., Zangerl, A. R., Garbutt, K. and Bazzaz, F. A. (1986). Within and between species variation in response to environmental gradients in *Polygonum pensylvanicum* and *Polygonum virginianum. Oecologia* **68**, 606–610.

Levin, D. A. (1981). Dispersal versus gene flow in plants. *Ann. Mo. Bot. Gard.* **68**, 233–253.

Levin, D. A. (1984). Immigration in plants: an exercise in the subjunctive. *In* "Perspectives in Plant Population Ecology" (R. Dirzo and J. Sarukhan, eds), pp. 242–260. Sinauer Associates Inc., Sunderland, MA.

Levin, D. A. and Kerster, H. W. (1971). Neighborhood structure in plants under diverse reproductive methods. *Am. Nat.* **105**, 345–354.

Levin, D. A. and Kerster, H. W. (1974). Gene flow in seed plants. *Evol. Biol.* **7**, 139–220.

Levins, R. (1968). "Evolution in a Changing Environment". Princeton University Press, Princeton.

Lewontin, R. C. (1957). The adaptation of populations to varying environments. *Cold Spring Harbor Symp. Quant. Biol.* **22**, 395–408.

Lewontin, R. C. (1974). "The Genetic Basis of Evolutionary Change". Columbia University Press, New York and London.

Linhart, Y. B. (1974). Intrapopulation differentiation in annual plants. I. *Veronica peregrina* raised under non-competitive conditions. *Evolution* **28**, 232–243.

Loucks, O. (1970). Evolution of diversity, efficiency and community stability, *Am. Zool.* **10**, 17–25.

Loveless, M. D. and Hamrick, J. L. (1984). Ecological determinants of genetic structure in plant populations. *Ann. Rev. Ecol. Syst.* **15**, 65–95.

Marshall, C. and Sagar, G. R. (1968). The distribution of assimilates in *Lolium multiflorum* Lam. following differential defoliation. *Ann. Bot.* **32**, 715–719.

McKey, D. (1979). The distribution of secondary compounds within plants. *In* "Herbivores: Their Interaction with Secondary Plant Metabolites" (G. A. Rosenthal and D. H. Janzen, eds), pp. 56–133. Academic Press, New York.

McNeilly, T. and Antonovics, J. (1968). Evolution in closely adjacent plant populations. IV. Barriers to gene flow. *Heredity* **23**, 205–218.

Mooney, H. A. and Gulmon, S. L. (1979). Environmental and evolutionary constraints on the photosynthetic characteristics of higher plants. *In* "Topics in Plant Population Biology" (O. Solbrig, S. K. Jain *et al.*, eds), pp. 316–337. Columbia University Press, New York.

Orzack, S. H. (1985). Population dynamics in variable environments. V. The genetics of homeostasis revisited. *Am. Nat.* **125**, 550–572.

Paine, R. T. (1966). Food web complexity and species diversity. *Am. Nat.* **100**, 65–75.

Parrish, J. A. D. and Bazzaz, F. A. (1979). Difference in pollination niche relationship in early and late successional plant communities. *Ecology* **60**, 597–610.

Parrish, J. A. D. and Bazzaz, F. A. (1982a). Responses of plants from three successional communities to a nutrient gradient. *J. Ecol.* **70**, 233–248.

Parrish, J. A. D. and Bazzaz, F. A. (1982b). Niche responses of early and late successional tree seedlings on three resource gradients. *Bull. Torrey Bot. Club* **109**, 451–456.

Parrish, J. A. D. and Bazzaz, F. A. (1985a). Nutrient content of *Abutilon theophrastii* seeds and the competitive ability of the resulting plants. *Oecologia* **65**, 247–251.
Parrish, J. A. D. and Bazzaz, F. A. (1985b). Ontogenetic niche shifts in old-field annuals. *Ecology* **66**, 1296–1302.
Pickett, S. T. A. and Bazzaz, F. A. (1978). Organization of an assemblage of early successional species on a soil moisture gradient. *Ecology* **59**, 1248–1255.
Primack, R. B. and Antonovics, J. (1981). Experimental ecological genetics in *Plantago*. V. Components of seed yield in the ribwort plantain *P. lanceolata* L. *Evolution* **35**, 1069–1079.
Regehr, D. L. and Bazzaz, F. A. (1979). The population dynamics of *Erigeron canadensis*, a successional winter annual. *J. Ecol.* **67**, 923–933.
Rhoades, D. F. (1983). Herbivore population dynamics and plant chemistry. *In* "Variable Plants and Herbivores in Natural and Managed Systems" (R. F. Denno and M. S. McClure, eds), Ch. 6. Academic Press, New York.
Rice, K. and Jain, S. (1985). Plant population genetics and evolution in disturbed environments. *In* "The Ecology of Natural Disturbance and Patch Dynamics" (S. T. A. Pickett and P. S. White, eds), pp. 287–303. Academic Press, New York.
Roach, D. A. (1986). Timing of seed production and dispersal in *Geranium carolinianum*: effects on fitness. *Ecology* **67**, 572–576.
Ryan, C. A. (1979). Proteinase inhibitors. *In* "Herbivores: Their Interaction with Secondary Plant Metabolites" (G. A. Rosenthal and D. H. Janzen, eds), pp. 559–618. Academic Press, New York.
Ryan, C. A. (1983). Insect-induced chemical signals regulating natural plant protection responses. *In* "Variable Plants and Herbivores in Natural and Managed Systems" (R. F. Denno and M. S. McClure, eds), Ch. 2. Academic Press, New York.
Schaal, B. A. (1984). Life-history variation, natural selection, and maternal effects in plant populations. *In* "Perspectives in Plant Population Ecology" (R. Dirzo and J. Sarukhan, eds), pp. 188–206. Sinauer Associates Inc., Sunderland, MA.
Schaal, B. A. (1985). Genetic variation in plant populations: from demography to DNA. *In* "Structure and Functioning of Plant Populations. 2. Phenotypic and Genotypic Variation in Plant Populations" (J. Haeck and J. W. Woldendorp, eds), pp. 321–341. North-Holland, Amsterdam.
Schlichting, C. (1986). The evolution of phenotypic plasticity in plants. *Ann. Rev. Ecol. Syst.* **17**, 667–693.
Schmid, B. A. (1985). Clonal growth in grassland perennials. III. Genetic variation and plasticity between and within populations of *Bellis perennis* and *Prunella vulgaris*. *J. Ecol.* **73**, 819–830.
Schmid, B. A. (1986). Spatial dynamics and integration within clones of grassland perennials with different growth form. *Proc. Roy. Soc. Lond.* **B228**, 173–186.
Schneller, J. J. and Schmid, B. S. (1982). Investigations on the intraspecific variability in *Athyrium filix-femina* (L.) Roth. *Bull. Mus. Natl Hist. Nat. Paris*, 4th Ser., Sect. B, Adansonia **3–4**, 215–228.
Schwaegerle, K. E. and Bazzaz, F. A. (1987). Differentiation among nine populations of *Phlox*. II. Response to environmental gradients. *Ecology* **68**, 54–64.
Schwaegerle, K., Garbutt, K. and Bazzaz, F. A. (1986). Differentiation among nine populations of *Phlox*. I. Electrophoretic and quantitative variation. *Evolution* **40**, 506–517.
Smith, W. H. (1976). Character and significance of forest tree root exudates. *Ecology* **57**, 324–331.
Snaydon, R. W. and Davies, M. S. (1972). Rapid population differentiation in a mosaic environment. I. Morphological variation in *Anthoxanthum odoratum*. *Evolution* **26**, 390–405.

Solbrig, O. T. (1981). Studies on the population biology of the genus *Viola*. II. The effect of plant size on fitness on *Viola sororia*. *Evolution* **35**, 1080–1093.

Solbrig, O. T. and Solbrig, D. J. (1984). Size inequalities and fitness in plant populations. *Oxford Surv. Evol. Biol.* **1**, 141–159.

Stanton, M. L. (1984). Seed variation in wild radish: effect of seed size on components of seedling and adult fitness. *Ecology* **65**, 1105–1112.

Sultan, S. E. (1987). Evolutionary implications of phenotypic plasticity in plants. *Evol. Biol.* **21**, 127–176.

Templeton, A. R. and Levin, D. A. (1979). Evolutionary consequences of seed pools. *Am. Nat.* **114**, 232–249.

Tilman, D. (1982). "Resource Competition and Community Structure". Princeton University Press, Princeton.

Turkington, R. and Harper, J. L. (1979). The growth, distribution and neighbor relationships of *Trifolium repens* in a permanent pasture. IV. Fine-scale biotic differentiation. *J. Ecol.* **67**, 245–254.

Venable, D. L. (1984). Using intraspecific variation to study the ecological significance and evolution of plant life-histories. *In* "Perspectives in Plant Population Ecology" (R. Dirzo and J. Sarukhan, eds), pp. 166–187. Sinauer Associates Inc., Sunderland, MA.

Whitham, T. G. (1983). Host manipulation of parasites: within-plant variation as a defense against rapidly evolving pests. *In* "Variable Plants and Herbivores in Natural and Managed Systems" (R. F. Denno and M. S. McClure, eds), pp. 15–41. Academic Press, London and New York.

Whittaker, R. H. (1975). "Communities and Ecosystems". Macmillan, New York.

Wright, S. (1931). Evolution in Mendelian populations. *Genetics* **16**, 97–159.

Zangerl, A. R. and Bazzaz, F. A. (1983). Plasticity and genotypic variation in photosynthetic behavior of an early and a late successional species of *Polygonum*. *Oecologia* **57**, 270–273.

Zangerl, A. R. and Bazzaz, F. A. (1984a). Effects of short-term selection along environmental gradients on variation in populations of *Amaranthus retroflexus* and *Abutilon theophrastii*. *Ecology* **65**, 207–217.

Zangerl, A. R. and Bazzaz, F. A. (1984b). Niche partitioning between two phosphoglucoisomerase genotypes in *Amaranthus retroflexus*. *Ecology* **65**, 218–222.

5 Complex Patterns of Genetic Differentiation and Phenotypic Plasticity versus an Outmoded Ecotype Terminology

JAMES A. QUINN

Department of Biological Sciences, Rutgers University, Piscataway, NJ, U.S.A.

I. Introduction

The "ecotype" was originally envisioned by Turesson (1922a) as the "ecological unit to cover the product arising as a result of the genotypical response of an ecospecies to a particular habitat." Turesson (1922b) proposed the more inclusive "ecospecies" "to cover the Linnean species or genotype compounds as they are realized in nature" and "ecophene" for "the reaction-types of the ecotypes arising through the modificatory influences of the combinations of extreme habitat factors given in nature." An important result of Turesson's early research (1922a,b, 1925, 1927, 1929, 1930) and a massive accumulation of evidence by other workers (Quinn, 1978) was to demonstrate conclusively that plant species in general show genetically based ecological differentiation.

Until the late 1950s, "ecotype" had primarily been used to refer to a grouping of populations in relation to a type of habitat or climate (Quinn, 1978). Although Turesson's definition (". . . genotypical response . . . to a particular habitat") is literally equivalent to a local population, he made his intentions clear by proposing a system designating ecotypes by the habitat-type they occupied. The genecological research of the Carnegie group (e.g., Clausen *et al.*, 1940; Clausen and Hiesey, 1958) reinforced this usage with the grouping of local populations of *Potentilla glandulosa* into climatic races or ecotypes corresponding to the subspecies *typica, reflexa, hanseni*, and *nevadensis* (Quinn, 1978). *Achillea*

Differentiation Patterns in Higher Plants
ISBN: 0-12-709425-3

was also studied along this central California seacoast-to-High Sierra transect and reported by Clausen and colleagues (1948) to exhibit a sequence of climatic races or ecotypes.

Today, the "ecotype" has no consistent usage. Not only is the term being used to indicate almost any degree or type of genetic difference below the species level, even for a group of individuals of the same genotype, but it has also been used for a group of morphologically similar races of different species (Böcher, 1977) and for an abstract concept of an ecosystem type (Phillips, 1971). The underlying causes for the confusing uses of "ecotype" are both practical and conceptual. It is experimentally impossible to delineate an ecotype unless every population of the species is studied, and our increased knowledge of genecology and population differentiation has made it clear that a Turesson ecotype may not be the best ecological unit and is definitely not an evolutionary unit (Quinn, 1978).

Heywood (1959) and Langlet (1971) predicted the eventual abandonment of the term "ecotype", and Quinn (1978) marshalled further arguments against its continued use. However, in contrast to its associated terms ("ecospecies", "coenospecies", and "ecophene") which are no longer in general use, "ecotype" continues to be used frequently, especially in ecological journals. The objective of this chapter will be to evaluate the implications of "individualistic" populations and of genetic differentiation in phenotypic plasticity to population biology and to the usefulness of the term "ecotype".

In this chapter the term "population" will be used to indicate a local population of one species — a number of individuals occurring together both temporally and spatially. A population is delimited by specifying a particular site which may be of any size, ranging from a square metre or less to a hectare or more. "Genetic differentiation" will refer to genetically based variation that is environmental or habitat correlated. "Phenotypic plasticity" will refer to the amount by which the expressions of individual characteristics of a genotype are changed by different environments (Bradshaw, 1965).

II. The "Individualistic" Population

A fundamental problem is that the Turesson ecotype assumes a lack of ecologically significant genetic variability among the genotypes and populations of an ecotype. Research in the late 1950s and the early 1960s produced data emphasizing the population as *the* evolutionary unit (e.g., Bradshaw, 1959; Jain and Bradshaw, 1966). These studies also indicated that effective population size should be measured in metres and not kilometres (Bradshaw, 1972) and emphasized the uniqueness of the local population in its particular morphological and physiological attributes; hence, the ecological similarity of the populations comprising a Turesson ecotype would depend on the degree of similarity of all the physical and biotic components of the local environment, assuming initially similar gene pools (Quinn, 1978).

Each combination of gene pool and environmental conditions is potentially unique, giving rise to a population potentially unpredictable in certain of its properties. Hume (1979), in a wide-ranging study of geographic variation in *Rumex crispus*, reported that "each parental and offspring population, grown in its natural habitat and under uniform conditions respectively, had a set of features distinguishing it from all other populations." Ten populations of *Danthonia caespitosa*, a perennial tussock grass of southern Australia, were found to be markedly unique in phenology, morphology, and physiology under natural habitats and controlled environments (Hodgkinson and Quinn, 1976, 1978; Quinn and Hodgkinson, 1983, 1984), even though several of the populations were closely adjacent (within 50 km) and/or occurred in similar habitats. However, even if the prevailing climatic conditions, topography, soils, and disturbance regime had been identical for *Danthonia* at all sites, populations of this sexually-reproducing perennial would have probably differed in ecologically important traits. The number and genetic composition of the initial colonists would never be the same, and this original gene pool will determine how (or if) the population can respond to selection pressures (Quinn and Hodgkinson, 1984). If it is assumed, however, that the gene pools were initially the same at every site, the character variability patterns of perennial plant populations could still be affected by the community species composition and density prevailing at the time of establishment of the majority of the population, temporal environmental differences that affect cohort survivorship curves (and the genetic composition of the survivors) in different ways at similar sites, and local differentiation as a result of competitive interactions (Turkington and Aarssen, 1984). Quinn and Hodgkinson (1984) also pointed out that a perennial plant population may consist of representatives from many generations or cohorts of various ages, each faced during their establishment phase with somewhat different selection regimes favouring different alleles or genotypes (Mulcahy, 1975; Hamrick *et al.*, 1979; Ledig and Korbobo, 1983). Therefore, unless a species is primarily asexual or highly inbreeding (but see Urbanska, 1985), each particular combination of gene pool and environmental conditions is unique, giving rise to a population unique or unpredictable in certain of its properties.

Populations may sometimes appear to be similar if they are compared only by measurements at specific developmental stages or by looking at the outcome of an adaptive response determined by the interaction of several characters. For example, in a common garden the three northernmost populations of *Danthonia caespitosa* had the greater (and amazingly comparable) number and weight of seeds per 10 g of biomass (Quinn and Hodgkinson, 1984), but these three populations achieved this similar response through very different allocation and developmental tactics (Table 1).

TABLE 1

Ranking (1 = highest) of the three northern populations of *Danthonia caespitosa* in characters determining the number (first three) and the weight (all four) of seeds produced per 10 g of shoot biomass in the common garden near Deniliquin, New South Wales, Australia (from Quinn and Hodgkinson, 1984).

Population	No. panicles per 10 g	No. spikelets per panicle	No. seeds per spikelet	Mean weight per seed
Nyngan	1	2–3	2	3
Lake Cargelligo	2	2–3	1	1
Deniliquin	3	1	3	2

III. Genetic Differentiation in Phenotypic Plasticity and its Implications

In his classic review, Bradshaw (1965) defined and documented the existence and importance of genetically based phenotypic plasticity in plant populations. Although he presented considerable evidence for populations differing in the amount and type of plasticity, Bradshaw noted the need for further experimental evidence of genetic control and natural selection for phenotypic plasticity. In 1974, Bradshaw summarized additional evidence and concluded that phenotypic plasticity can be:

"(a) specific for individual characters
(b) specific in relation to particular environmental influences
(c) specific in pattern and direction
(d) under genetic control not necessarily related to heterozygosity."

There is a need for further research into the amount of genetic variability in phenotypic plasticity in natural populations (Schlichting, 1986) and recent research has provided a range of interesting examples of genetic differentiation in phenotypic plasticity in relation to local environmental variability and predictability, only a few of which can be briefly mentioned here. In Colorado, progenies of the annual *Collomia linearis* from a disturbed site were phenotypically more plastic than those from an adjacent meadow (Wilken, 1977); an analysis of nine morphological characters in progenies subjected to 36 different environmental conditions showed the among-environment variance components to be higher in the disturbed population for all nine (six were significantly greater). In New Jersey, genotypes from successional populations of *Andropogon scoparius*, a perennial grass, were compared in height growth, phenology, and shoot biomass allocation under field and two greenhouse conditions (Roos and Quinn, 1977). The oldest population (40 yr) showed the most phenotypic plasticity, not only for reproductive effort (Table 2) but also for the other characters; it was hypothesized that over several decades periodic droughts on these relatively shallow soils and seasonally variable amounts of shading and competition had "screened out" those genotypes

TABLE 2

Percentage reproductive effort for eight genotypes of each of three populations of *Andropogon scoparius* determined first in the field and subsequently in the greenhouse under reduced light intensities. Means within a row or column followed by the same letter are not significantly different at the 0.05 level (from Roos and Quinn, 1977)

Population and age	Percentage reproductive effort		
	Field	Greenhouse (shaded)	Greenhouse (no shade)
CH-3	52 a	50 a	63 b
WP-10	45 a	46 ac	55 c
JFK-40	18 b	43 a	55 c

showing less developmental or physiological plasticity. Two populations of *Heliotropium curassavicum*, a succulent annual or short-lived perennial, one from sand dunes facing the sea at Point Reyes, California, and the other from the floor of Death Valley, were found to differ in their photosynthetic and stomatal responses to desiccation (Roy and Mooney, 1982). The desert plants had a plastic stomatal response to dry air growing conditions which resulted in a greater photosynthetic performance at negative water potentials; the capacity of Death Valley plants to photosynthesize at low water potentials when grown at low humidity came from a decreased sensitivity of stomata to leaf xylem water potential. Thus, Roy and Mooney (1982) concluded that plants originating from a habitat with a wider range of water conditions had a greater acclimation potential. Bell and Quinn (in press) grew plants from six New Jersey populations of *Dichanthelium clandestinum*, a perennial grass with chasmogamous and cleistogamous (CL) flowers on the same individual, along a soil moisture gradient under controlled conditions. The population responses formed a continuum from no difference in percentage biomass allocation to CL along the soil moisture gradient to a sharp drop in CL allocation at the lower end of the gradient to a relatively steady decrease in CL allocation with decreasing soil moisture. The two populations whose plants had a relatively high allocation to CL reproduction at low soil moisture were also the two populations which had the lowest soil moisture in the field. My last example involves the growth of two *Heterotheca subaxillaris* populations from seeds under two levels of nitrate availability (Mihaliak, 1986). A greater response to nitrate availability was observed among plants from an abandoned old field which had received periodic fertilization than from the individuals of a low-nutrient coastal sand dune population. The results thus indicated less phenotypic plasticity among the individuals in the coastal population, where nitrate availability had been uniformly low.

A. Relation between genetic variation and phenotypic plasticity

It has been postulated that genetic variation and phenotypic plasticity are alternative strategies (Lewontin, 1957; Levins, 1963; Marshall and Jain, 1968; Jain, 1978, 1979), i.e., that selection will operate to maximize either genetic variation or phenotypic plasticity (Scheiner and Goodnight, 1984). This hypothesis has been supported mostly by theoretical arguments and the demonstration of a negative correlation between genetic polymorphism and plasticity in three pairs of congeneric species (Marshall and Jain, 1968; Jain, 1969, 1979). More recently, most investigators have found either the opposite relationship (Wilken, 1977) or no relationship (Gottlieb, 1977; Scheiner and Goodnight, 1984; Schlichting and Levin, 1984; but see Silander, 1985).

Phenotypic plasticity depends on regulatory genes and alternative enzyme systems or metabolic pathways, triggered by specific environmental stimuli, and therefore may be enhanced by genetic variation. It is not surprising that Professor H. Hurka (pers. comm., 1986), following extensive studies of the genetic and phenotypic variation in *Capsella bursa-pastoris* populations in Europe, has found no consistent relation between the amount of genetic variation and the amount of phenotypic plasticity, and that sometimes a greater amount of genetic variation means more plasticity. There is also no reason to suspect that selection for genetic variation and phenotypic plasticity would always be mutually exclusive. As Scheiner and Goodnight (1984) suggested, long-term cyclic fluctuation could select for genetic tracking, and short-term random fluctuations could select for phenotypic plasticity. In addition, it must be remembered that different characters may be under differential selection pressures in the same environment, leading to fixation for some and polymorphism or plasticity in others.

B. Relation of phenotypic plasticity to the character means or responses used to define ecotypes

Defining ecotypes on the basis of physiological or morphological character means has usually ignored genetic differentiation in phenotypic plasticity. There is a striking exception in which the ability of the photosynthetic apparatus to adjust to light intensity was used to delineate "ecotypes" of *Solidago virgaurea* from exposed and shaded habitats (Björkman and Holmgren, 1963). However, other characters were not utilized, and, for example, the two populations of exposed habitats probably differed profoundly in their growth responses to temperature, photoperiod, duration of the growing season, etc., as one came from a lowland area in southern Sweden and the other from the alpine tundra in northernmost Norway.

Phenotypic plasticity may vary independently of character means such that populations with similar means may differ in the amount and pattern of plasticity across a range of environments (Schlichting and Levin, 1986). Khoshoo (1958) grew the diploid *Sisymbrium irio* (Brassicaceae) and the teraploids *S. irio caulis* and *S. irio subcaulis* in four garden environments combining sun and shade and moist and dry conditions, and his data were recently analysed for the relative amounts and patterns of phenotypic plasticity by Schlichting and Levin (1986). Three characters for which the populations showed almost no variation in means differed markedly in the amounts of phenotypic plasticity (CVs) among populations, and one of them also differed in the patterns of plasticity (ρ) shown by the populations (Table 3).

TABLE 3

Character means and amounts (CVs) and patterns (ρ) of phenotypic plasticity in *Sisymbrium* (from Table 1 of Schlichting and Levin, 1986).

Character	Mean			Coefficient of variation			Spearman correlation (ρ) between taxa		
	D[a]	C[a]	S[a]	D	C	S	D–C	C–S	D–S
Height (log)	3.9	3.6	3.9	34	45	8	1.00	0.80	0.80
Leaf length (log)	2.0	1.7	2.4	64	89	27	1.00	1.00	1.00
Leaf width (log)	1.8	1.4	2.1	77	130	35	1.00	1.00	1.00
No. of branches	6.5	3.2	2.5	99	93	33	0.80	0.95	0.63
Siliqua length	38.2	33.2	34.0	16	13	4	0.95	1.00	0.95
No. of seeds per siliqua	62.8	55.0	60.5	22	28	6	0.95	–0.06	–0.32

[a]D = diploid; C = caulis; S = subcaulis.

Progenies of wild populations of *Capsella bursa-pastoris* along a transect from northern to southern Europe showed pronounced intraspecific variation in developmental time until flowering (Neuffer and Hurka, 1986). In addition, the populations were observed to vary in their amount and pattern of plasticity. Figure 1 shows the mean and range of the date of first flowering during different years for four Swiss populations in open-field random-block experiments. Within any given year, there are obvious differences between both the population means and the total flowering range. Since the environment for a given year was identical for all populations, differences among plants in the same year were attributed to genetic variation, and variation in flowering behaviour from year to year was attributed to phenotypic plasticity (Neuffer and Hurka, 1986). Populations with the same mean in a given year may vary that year in range and differ in the other years in both mean and range (e.g., 279 and 281 or 280 and 281). This annual variability is even more evident from the values of the coefficients of variation, CVs, for these progenies raised from genetically similar seed samples in different years (Table 4).

Phenotypic plasticity has been said to diminish a character's taxonomic utility because the environmental contribution to phenotypic variation obscures

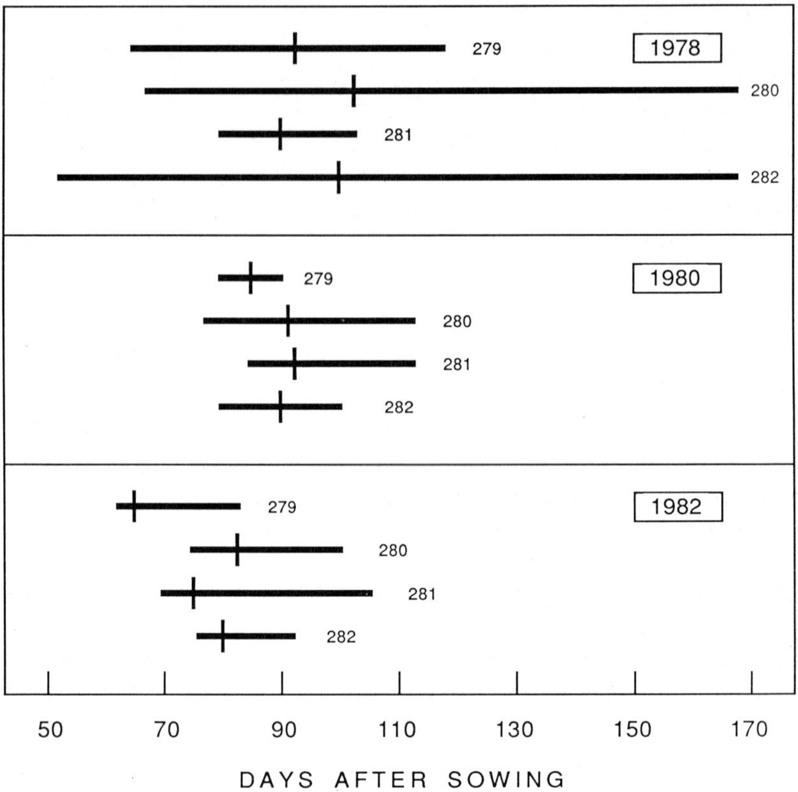

FIG. 1 Onset of flowering (days after sowing) in 1978, 1980, and 1982 of plants from four Switzerland populations of *Capsella bursa-pastoris* grown from genetically similar seed samples each year at the field station of the University of Münster, FRG. Total ranges and population means are given (extracted from Fig. 1 of Neuffer and Hurka, 1986).

TABLE 4

Coefficients of variation for the onset of flowering for four Switzerland populations of *Capsella bursa-pastoris* in different years at the Münster field station (extracted from Table 2 of Neuffer and Hurka, 1986).

Population	Coefficients of variation		
	1978	1980	1982
279	0.238	0.058	0.090
280	0.278	0.124	0.090
281	0.161	0.093	0.125
282	0.239	0.072	0.082

genetically based relationships (Davis, 1983). To ignore phenotypic plasticity in ecological differentiation and taxonomic research may itself obscure important adaptive responses and genetically based relationships. Schlichting and Levin (1986) documented instances of substantial differentiation between closely related taxa for phenotypic plasticity and suggested that evolutionary shifts in plasticity have played an important, but underestimated, role in the differentiation of taxa. Considering its adaptive importance in plants, Morisset and Boutin (1984) argued that the phenomenon of phenotypic plasticity should become a central theme in biosystematics.

C. Relation of the detection of differential population responses to the populations and the environments utilized

Both the detection of differential population responses to a common environment and the variability among populations in their responses to a range of environments depend on which common environments are utilized, their number, their appropriateness to the populations being studied, and the populations and the characters chosen for analysis. Although the preceding is "obvious" to any population biologist, it is not obvious from the literature that everyone takes it seriously. It is also not possible to judge the degree of misinformation arising from poorly designed research. However, by looking at research which has included a considerable number of genotypes and populations under a wide range of environments, we can ask "what if" questions such as:

(1) If only two or three populations had been utilized, would the results have been different?
(2) Would estimates of the relative amounts of plasticity depend on the environments utilized?
(3) Can field results be used to predict responses under controlled environments, or vice versa? and
(4) Do different characters show similar patterns of variability, and are these patterns predictable?

To answer the above questions for *Danthonia caespitosa* in southern Australia, research that included data on 21 characters in nine populations under field, transplant garden glasshouse, and phytotron conditions (Quinn and Hodgkinson, 1984) will be utilized. First, if information from only two or three populations had been used, it would have frequently been highly misleading! One example would be panicle weights — if only Deniliquin, Heathcote, and Conara Junction had been sampled for panicle weights under a common environment (Table 5), there would have been highly significant population differences correlated perfectly with latitude, rather than the marked variation without latitudinal pattern shown in both field and garden (Quinn and Hodgkinson, 1984). Another example would be seed weights (Tables 6 and 7), for which the nine populations showed significant

TABLE 5

Panicle weights and shoot biomass allocation to panicles for five populations of *Danthonia caespitosa* in the common garden near Deniliquin[a] (from Quinn and Hodgkinson, 1984).

Population (from N to S)	Panicle weight (mg) (n = 35)	Panicles as percentage of shoot weight (n = 10)
Nyngan	101.6 a	31.4 a
Lake Cargelligo	144.9	39.3
Deniliquin	219.7	31.7 a
Heathcote	99.5 a	17.3 b
Conara Junction	69.2	12.8 b

[a]Means within a column followed by the same letter are not significantly different at the 0.05 level.

differences but no discernible selective cause for the variability exhibited. If we had collected data only from Lake Cargelligo and Heathcote, or only from Lake Cargelligo, Deniliquin, and Conara Junction, the results would have agreed perfectly with past data and theory that predict larger seeds for more rapid germination and successful establishment in xeric sites (Quinn *et al.*, 1972), or larger seeds for success where both growing season and period for seedling establishment are shorter (McWilliams *et al.*, 1968).

In response to the second "what if" question, the estimates of the relative amounts of population plasticity for certain characters did depend on the environments utilized. In the case of seed weights, the field weights were always significantly different from the garden weights, higher for two populations and lower for two others (Table 6). For the five populations grown in the phytotron, their seed weights under three temperature regimes either accentuated or muted the perception of plasticity (Table 7). The two populations with seeds at 36/31°C produced light seeds that greatly amplified the range of their seed weights, and Conara Junction added weights of 0.284 mg (24/19°C) and 0.179 mg (30/25°C) to a garden weight of 0.323 mg. On the other hand, Deniliquin added weights of 0.306 mg (24/19°C) and 0.317 mg (30/25°C) to its field weights of 0.302 mg (1972) and 0.294 mg (1980), so that if the garden environment had not been utilized, Deniliquin would have demonstrated no significant plasticity in seed weight (Tables 6 and 7).

Thirdly, our results indicated that field results can not be used to predict responses under controlled environments, or vice versa. For example, the population rankings from field studies erroneously predicted comparative population responses to a common environment in more than 50% of the cases (Quinn and Hodgkinson, 1984).

The answer to the fourth "what if" question is that phenotypic plasticity differed among the populations, and the various characters showed strikingly different patterns of variability among populations (Quinn and Hodgkinson, 1984).

TABLE 6

Mean weight per seed (mg) for *Danthonia caespitosa* populations at the field sites in 1972 and 1980 and at the common garden area near Deniliquin[a] (from Quinn and Hodgkinson, 1984).

Population (from N to S)	Field 1972	Field 1980	Common garden area
Nyngan	—	0.309 bc	0.282 a
Lake Cargelligo	—	0.329 a	0.397
Hay	—	0.346 a	—
Deniliquin	0.302 A	0.294 dA	0.377
Moama	0.282 b	0.298 cd	—
Elmore	0.342 a	0.584	—
Heathcote	0.334 aA	0.322 abA	0.285 a
Tooradin	0.271 b	0.249	—
Conara Junction	—	—	0.323

[a]Values are means of 20 lots of 25 oven-dried seeds. Means followed by the same letter (upper case for row and lower case for column) are not significantly different at the 0.05 level.

TABLE 7

Mean weight per seed (mg) for *Danthonia caespitosa* populations in the Ceres Phytotron at three temperature regimes[a] (from Quinn and Hodgkinson, 1984).

Population (from N to S)	Temperature regime 24/19°C	30/25°C	36/31°C
Nyngan	0.290 abA	0.308 aA	0.229
Lake Cargelligo	0.436 A	0.424 A	0.260
Deniliquin	0.306 aA	0.317 aA	—
Heathcote	0.271 A	0.282 A	—
Conara Junction	0.284 b	0.179	—

[a]Values are means of 20 lots of 25 oven-dried seeds, except for Conara Junction at 30/25, Nyngan at 36/31, and Lake Cargelligo at 36/31, which were determined from 17, 13, and 9 lots, respectively. Only a few seeds were produced by the three southern populations at 36/31. Means followed by the same letter (upper case for row and lower case for column) are not significantly different at the 0.05 level.

Considerable evidence exists for other species that populations may differ in their plasticity for particular characters, that different characters may show different levels of plasticity, and that these characters may separately respond to specific environmental factors (Bradshaw, 1965, 1974; Schlichting, 1986).

An excellent example of the relative amount of plasticity or differential population responses being a function of the specific environments to which the populations are exposed is a recent study of the annual *Impatiens capensis* (Simpson *et al.*, 1985). One aspect of this extensive research involved the transplanting of seedlings from six populations to two of the population sites, enabling a

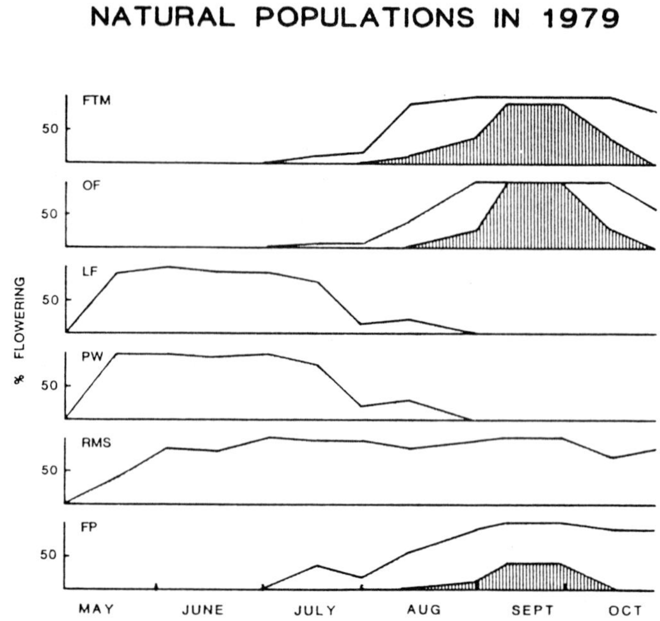

FIG. 2 Floral phenology for six *Impatiens capensis* populations germinating naturally in 1979 and as transplants to two sites in 1981, showing initiation of flowering and percentage of individuals in flower. The non-shaded area represents cleistogamous flowering and the shaded portion chasmogamous flowering (from Simpson *et al.*, 1985).

comparison of each population's flowering phenology and production of chasmogamous flowers at the home site in 1979 with that at the transplant sites in 1981 (Fig. 2). Initiation of flowering was relatively constant for a population and related to the length of the growing season at its site, suggesting genetically fixed differences among populations. In contrast, length of flowering period and percentage of plants producing chasmogamous flowers showed marked plasticity, and the two 1981 environments had differing effects on these population responses.

A final point that deserves attention is that phenotypic differences among sites may be population-specific adaptive products of the local genotype–environment interaction, disappearing under controlled or common environments. To illustrate this possibility, our research on seaside goldenrod (*Solidago sempervirens*), a perennial of sand dune and saline areas along the Atlantic Coast of the United States, will be utilized (Cartica and Quinn, 1980, 1982). Four populations were studied along an 0.8 km transect across the barrier beach at Island Beach State Park, New Jersey (Fig. 3), and the population sites differed in salt spray deposition, vegetation densities, and frequency of major and minor disturbance (Cartica and Quinn, 1980, 1982). Our hypothesis was that selection pressures had resulted in genetically fixed differences in growth and biomass allocation among these closely adjacent populations. Percentage reproductive effort in the field ranged

TRANSPLANTS TO FP IN 1981

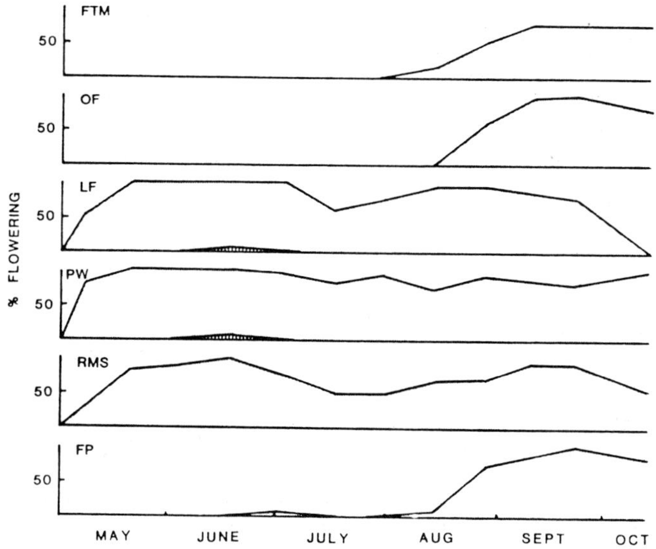

FIG. 2 (Continued).

TRANSPLANTS TO LF IN 1981

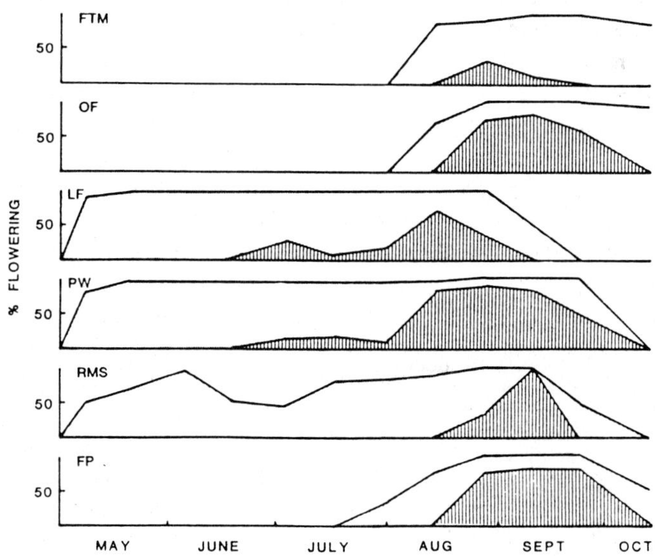

FIG. 2 (Continued).

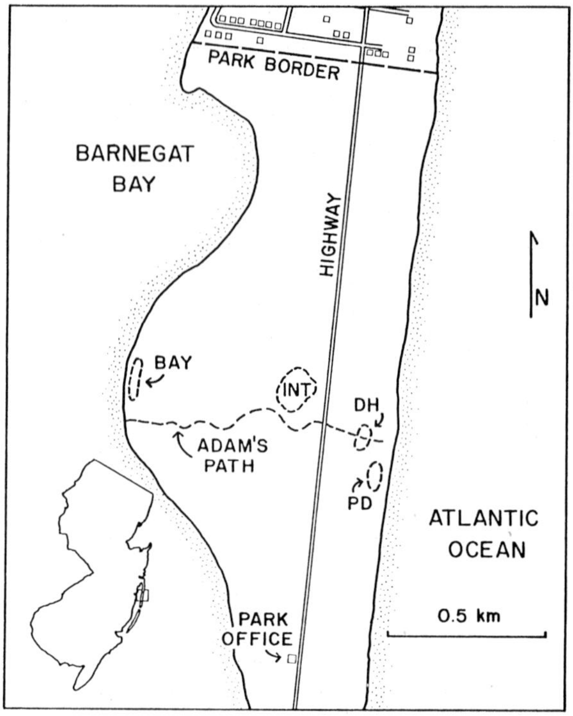

FIG. 3 Approximate location of the four *Solidago sempervirens* population sites at Island Beach State Park, New Jersey (from Cartica and Quinn, 1980).

from 14 to 28%, and there were significant differences among populations in allocation to component parts (Fig. 4). BAY plants were taller and heavier and had the greatest reproductive effort. INT, the only population with even a partial tree cover, was lowest in weight and reproductive effort. The high allocation to leaves at INT appeared to be a response to shading. In general, those populations (BAY and PD) that occurred in the most unstable sites, subjected to frequent catastrophic events, had a greater reproductive effort and fecundity than those populations (DH and INT) that occurred in the more stable sites (Cartica and Quinn, 1982). To determine whether these population differences were genetically fixed or environmentally induced, pieces (ramets) of field genotypes were grown in our greenhouse for 15 months. There were no significant population differences in height, biomass, resource allocation (Fig. 4), or fecundity. Did we thus conclusively demonstrate that there were no genetic differences in resource allocation among those four populations? Absolutely not! Those field differences were the phenotypic products in each case of the local genotype–environment interaction. Although the populations responded the same to their common greenhouse environment, we do *not* know if they would have responded in the same way

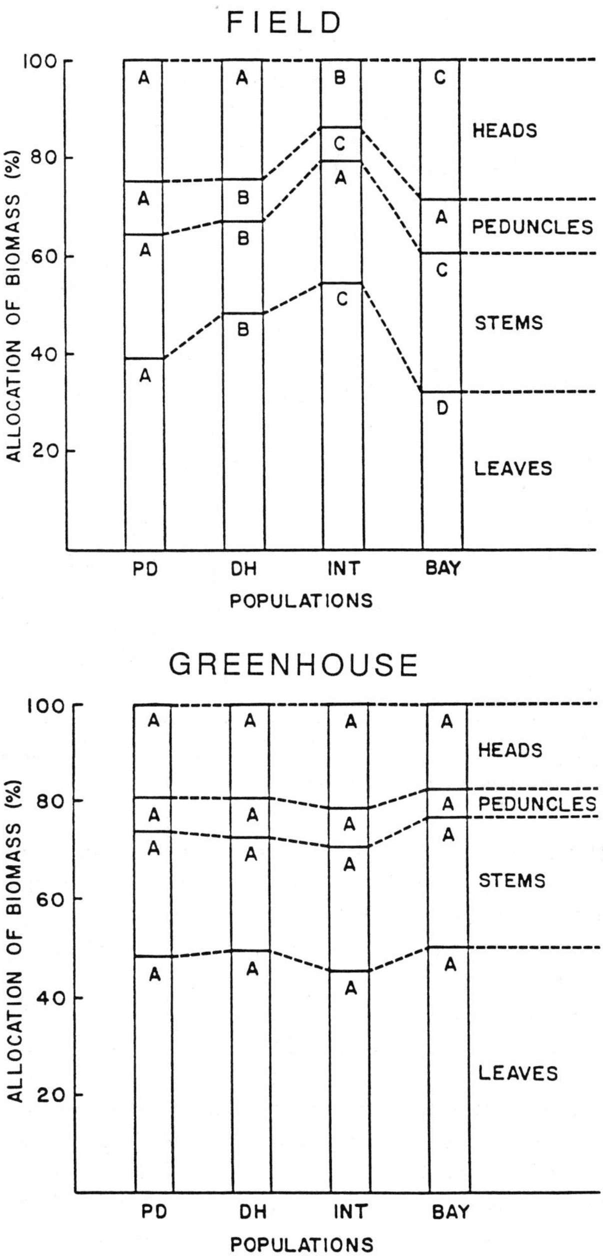

FIG. 4 Percentage resource allocation of *Solidago sempervirens* populations in the field and in the greenhouse. Values for the field represent means of 32–42 genotypes and those from the greenhouse are means of 5–8 genotypes. Values within any plant component containing the same letter are not significantly different at the 0.05 level (from Cartica and Quinn, 1982).

if they had been reciprocally transplanted to each site, i.e., all populations should be exposed to the BAY environment, or to the INT site, etc. However, due to the unique and highly vulnerable nature of the barrier beach vegetation at Island Beach State Park, we were not allowed to attempt such reciprocal transplanting. That the expression (and, therefore, detection) of genetic differences may vary from one transplant environment to another was frequently demonstrated in the Clausen, Keck, and Hiesey studies in the 1940s and 1950s (e.g., Clausen *et al.*, 1940, 1948; Clausen and Hiesey, 1958). Primack and Antonovics (1982) speculated that the large differences among field populations of *Plantago lanceolata* that mostly disappeared in the phytotron may have been the result of genotype–environment interactions that are highly environment specific. The basis for their statement was that the correlation of field and phytotron plants was greatest in the phytotron treatment with low nutrients, conditions most similar to those in the field.

IV. Conclusions

The use of the term "ecotype" creates both practical and conceptual problems in the identification and characterization of ecological and evolutionary units. Instead, there seems to be a general consensus today that populations are unique end-products of their genetic and ecological histories and that, as such, populations are important units of variability for the systematist, ecologist, and evolutionary biologist. It is also widely recognized that phenotypic plasticity is genetically determined and a product of selection, and must be considered by the systematist and the ecologist in assessing taxonomic relationships and ecological differentiation. Future research must bring these two ideas together more frequently in well-designed experiments. We have no further need for comparative population studies done only in the field or only under one controlled environment, or for studies utilizing only one population of a species or only two or three individuals each from a few populations. If we are to appreciate the range of adaptive variation in populations and species, we must not only compare their differential responses to a range of environments, but also equally emphasize the amounts and patterns of plasticity exhibited across this range of environments. Given the potentially unpredictable genetic variation in populations and its mostly unknown relationship with the amounts and patterns of plasticity, it is experimentally impossible to delineate an ecotype as either an ecological or evolutionary unit unless every population of the species is studied.

References

Bell, T. J. and Quinn, J. A. (In press). Effects of soil moisture and light intensity on the chasmogamous and cleistogamous components of reproductive effort of *Dichanthelium clandestinum* populations. *Can. J. Bot.*

Björkman, O. and Holmgren, P. (1963). Adaptability of the photosynthetic apparatus to light intensity in ecotypes from exposed and shaded habitats. *Physiol. Plant.* **16**, 889–914.

Böcher, T. W. (1977). Convergence as an evolutionary process. *Bot. J. Linn. Soc.* **75**, 1–19.

Bradshaw, A. D. (1959). Population differentiation in *Agrostis tenuis* Sibth. I. Morphological differentiation. *New Phytol.* **58**, 208–227.

Bradshaw, A. D. (1965). Evolutionary significance of phenotypic plasticity in plants. *Adv. Gen.* **13**, 115–155.

Bradshaw, A. D. (1972). Some of the evolutionary consequences of being a plant. *Evol. Biol.* **5**, 25–47.

Bradshaw, A. D. (1974). Environment and plasticity. *Brookhaven Symp. Biol.* **25**, 75–94.

Cartica, R. J. and Quinn, J. A. (1980). Responses of populations of *Solidago sempervirens* (Compositae) to salt spray across a barrier beach. *Am. J. Bot.* **67**, 1236–1242.

Cartica, R. J. and Quinn, J. A. (1982). Resource allocation and fecundity of populations of *Solidago sempervirens* along a coastal dune gradient. *Bull. Torrey Bot. Club* **109**, 299–305.

Clausen, J. and Hiesey, W. M. (1958). Experimental studies on the nature of species. IV. Genetic structure of ecological races. *Carnegie Inst. Wash. Publ.* **615**.

Clausen, J., Keck, D. D. and Hiesey, W. M. (1940). Experimental studies on the nature of species. I. Effect of varied environments on western North American plants. *Carnegie Inst. Wash. Publ.* **520**.

Clausen, J., Keck, D. D. and Hiesey, W. M. (1948). Experimental studies on the nature of species. III. Environmental responses of climatic races of *Achillea. Carnegie Inst. Wash. Publ.* **581**.

Davis, J. I. (1983). Phenotypic plasticity and the selection of taxonomic characters in *Puccinellia* (Poaceae). *Syst. Bot.* **8**, 341–353.

Gottlieb, L. D. (1977). Phenotypic variation in *Stephanomeria exigua* ssp. *coronaria* (Compositae) and its recent derivative species "Malheurensis". *Am. J. Bot.* **64**, 873–880.

Hamrick, J. L., Linhart, Y. B. and Mitton, J. B. (1979). Relationships between life history characteristics and electrophoretically detectable genetic variation in plants. *Ann. Rev. Ecol. Syst.* **10**, 173–200.

Heywood, V. H. (1959). The taxonomic treatment of ecotypic variation. *Systematics Ass. Publ.* **3**, 87–112.

Hodgkinson, K. C. and Quinn, J. A. (1976). Adaptive variability in the growth of *Danthonia caespitosa* Gaud. populations at different temperatures. *Aust. J. Bot.* **24**, 381–396.

Hodgkinson, K. C. and Quinn, J. A. (1978). Environmental and genetic control of reproduction in *Danthonia caespitosa* populations. *Aust. J. Bot.* **26**, 351–364.

Hume, L. (1979). Variation among populations of the wide-spread perennial weed, *Rumex crispus* L. Ph.D. thesis, University of Western Ontario, London, Ontario, Canada.

Jain, S. K. (1969). Comparative ecogenetics of two *Avena* species occurring in central California. *Evol. Biol.* **3**, 73–118.

Jain, S. K. (1978). Inheritance of phenotypic plasticity in soft chess, *Bromus mollis* L. (Gramineae). *Experientia* **34**, 835–836.

Jain, S. K. (1979). Adaptive strategies: polymorphism, plasticity, and homeostasis. *In* "Topics in Plant Population Biology" (O. T. Solbrig, S. K. Jain, G. B. Johnson and P. H. Raven, eds), pp. 160–187. Columbia University Press, New York.

Jain, S. K. and Bradshaw, A. D. (1966). Evolutionary divergence among adjacent plant populations. I. The evidence and its theoretical analysis. *Heredity* **21**, 407–441.

Khoshoo, T. N. (1958). Biosystematics of *Sisymbrium irio* complex. I. Modifications in phenotype. *Caryologia* **11**, 109–132.

Langlet, O. (1971). Two hundred years genecology. *Taxon* **20**, 653–722.

Ledig, F. T. and Korbobo, D. R. (1983). Adaptation of sugar maple populations along altitudinal gradients: photosynthesis, respiration, and specific leaf weight. *Am. J. Bot.* **70**, 256–265.
Levins, R. (1963). Theory of fitness in a heterogeneous environment. II. Developmental flexibility and niche selection. *Am. Nat.* **97**, 75–90.
Lewontin, R. C. (1957). The adaptations of populations to varying environments. *Cold Spring Harbor Symp. Quant. Biol.* **22**, 395–408.
Marshall, D. R. and Jain, S. K. (1968). Phenotypic plasticity of *Avena fatua* and *A. barbata. Am. Nat.* **102**, 457–467.
McWilliams, E. L., Landers, R. Q. and Mahlstede, J. P. (1968). Variation in seed weight and germination in populations of *Amaranthus retroflexus* L. *Ecology* **49**, 290–296.
Mihaliak, C. A. (1986). Differences in phenotypic plasticity between populations of *Heterotheca subaxillaris*: response to nitrate availability. *Am. J. Bot.* **73**, 672 (Abstr.).
Morisset, P. and Boutin, C. (1984). The biosystematic importance of phenotypic plasticity. *In* "Plant Biosystematics" (W. F. Grant, ed.), pp. 293–306. Academic Press Canada, Don Mills, Ontario.
Mulcahy, D. L. (1975). Differential mortality among cohorts in a population of *Acer saccharum* (Aceraceae) seedlings. *Am. J. Bot.* **62**, 422–426.
Neuffer, B. and Hurka, H. (1986). Variation of development time until flowering in natural populations of *Capsella bursa-pastoris* (Cruciferae). *Plant Syst. Evol.* **152**, 277–296.
Phillips, E. A. (1971). "Basic Ideas in Biology". MacMillan, New York.
Primack, R. B. and Antonovics, J. (1982). Experimental ecological genetics in *Plantago*. VII. Reproductive effort in populations of *P. lanceolata* L. *Evolution* **36**, 742–752.
Quinn, J. A. (1978). Plant ecotypes: ecological or evolutionary units? *Bull. Torrey Bot. Club* **105**, 58–64.
Quinn, J. A. and Hodgkinson, K. C. (1983). Population variability in *Danthonia caespitosa* (Gramineae) in responses to increasing density under three temperature regimes. *Am. J. Bot.* **70**, 1425–1431.
Quinn, J. A. and Hodgkinson, K. C. (1984). Plasticity and population differences in reproductive characters and resource allocation in *Danthonia caespitosa* (Gramineae). *Bull. Torrey Bot. Club* **111**, 19–27.
Quinn, J. A., Rotsettis, J. and Fairbrothers, D. E. (1972). Inflorescence characters and reproductive proficiency in *Danthonia sericea* populations. *Am. J. Bot.* **59**, 627–631.
Roos, F. H. and Quinn, J. A. (1977). Phenology and reproductive allocation in *Andropogon scoparius* (Gramineae) populations in communities of different successional stages. *Am. J. Bot.* **64**, 535–540.
Roy, J. and Mooney, H. A. (1982). Physiological adaptations and plasticity to water stress of coastal and desert populations of *Heliotropium curassavicum* L. *Oecologia (Berl.)* **52**, 370–375.
Scheiner, S. M. and Goodnight, C. J. (1984). The comparison of phenotypic plasticity and genetic variation in populations of the grass *Danthonia spicata. Evolution* **38**, 845–855.
Schlichting, C. D. (1986). The evolution of phenotypic plasticity in plants. *Ann. Rev. Ecol. Syst.* **17**, 667–693.
Schlichting, C. D. and Levin, D. A. (1984). Phenotypic plasticity of annual *Phlox*: tests of some hypotheses. *Am. J. Bot.* **71**, 252–260.
Schlichting, C. D. and Levin, D. A. (1986). Phenotypic plasticity: an evolving plant character. *Biol. J. Linn. Soc.* **29**, 37–47.
Silander, J. A. (1985). The genetic basis of the ecological amplitude of *Spartina patens*. II. Variance and correlation analysis. *Evolution* **39**, 1034–1052.
Simpson, R. L., Leck, M. A. and Parker, V. T. (1985). The comparative ecology of

Impatiens capensis Meerb. (Balsaminaceae) in central New Jersey. *Bull. Torrey Bot. Club* **112**, 295–311.
Turesson, G. (1922a). The species and the variety as ecological units. *Hereditas* **3**, 100–113.
Turesson. G. (1922b). The genotypical response of the plant species to the habitat. *Hereditas* **3**, 211–350.
Turesson, G. (1925). The plant species in relation to habitat and climate. *Hereditas* **6**, 147–236.
Turesson, G. (1927). Contributions to the genecology of glacial relics. *Hereditas* **9**, 81–101.
Turesson, G. (1929). Ecotypical selection in Siberian *Dactylis glomerata* L. *Hereditas* **12**, 335–351.
Turesson, G. (1930). The selective effect of climate upon the plant species. *Hereditas* **14**, 99–152.
Turkington, R. and Aarssen, L. W. (1984). Local-scale differentiation as a result of competitive interactions. *In* "Perspectives on Plant Population Ecology" (R. Dirzo and J. Sarukhan, eds), pp. 107–127. Sinauer Associates Inc., Sunderland, MA.
Urbanska, K. M. (1985). Some life history strategies and population structure in asexually reproducing plants. *Bot. Helv.* **95**, 81–97.
Wilken, D. H. (1977). Local differentiation for phenotypic plasticity in the annual *Collomia linearis* (Polemoniaceae). *Syst. Bot.* **2**, 99–108.

6 Hybridization and the Plant Species

CLIVE A. STACE

Department of Botany, University of Leicester, Leicester, U.K.

I. Introduction

Although anyone really familiar with wild or cultivated plants frequently encounters interspecific hybrids, many people in different but related fields (e.g. genetics, zoology) are surprised when shown the botanical facts, or even doubt them. Hybrids are certainly much rarer among animals, and their absence or sterility has justifiably led to the widespread use of the breeding barrier as a major criterion of species limits in both vertebrates and invertebrates. Exceptions occur, but they *are* exceptions. The temptation to extrapolate from animals to plants has been too great for some, but it is a route to taxonomic chaos. The broad facts about the extent of hybridization in at least higher plants are now too well documented for there to be any excuses for misapprehension about the situation in plants.

This chapter attempts to throw some light on the more important question of the *importance* of hybridization in terms of plant evolution, by summarizing some basic data, surveying known routes of speciation via hybridization, suggesting ways in which we might estimate the quantitative significance of these routes, and presenting two actual examples that illustrate some of the general points. Throughout the chapter the term hybrid will be used to designate a taxonomic hybrid, i.e. one between different taxa, but those considered are almost solely interspecific (or intergeneric) hybrids.

Differentiation Patterns in Higher Plants
ISBN: 0-12-709425-3

II. Frequency and Extent of Hybridization

The only quantitative estimate of the numbers of interspecific natural hybrids of which I am aware is that by Stace (1975a, 1975b) for the flora of the British Isles. In those works 626 hybrids within the vascular plants were reckoned to be well substantiated, and since that time (to the end of 1985) 91 further hybrids have been added. In addition, 122 combinations are represented by equivocal evidence. Allowing for 50% of the latter to be correct, the total stands at 778 interspecific vascular plant hybrids in a total flora (including established aliens, whose hybrids are included in the above figures, but excluding agamospecies in such notorious genera as *Rubus, Hieracium* and *Taraxacum*) of about 2500 species.

It is difficult to know how accurately one can extrapolate such figures to a world-wide scale. Firstly, we know little about the population biology of most tropical plants, and cannot judge whether hybrids there are more or less common than in the British Isles. Secondly, the degree of overlap between the figures for the British Isles and adjacent regions could give cause for debate. However, the above figures do *not* include the many (certainly over 450, perhaps substantially more) combinations involving two British species that are known elsewhere but not in the British Isles, let alone the still more that involve one British and one non-British species. Hence, our best guesses at a world total are obtained by simple multiplication of the British figures. The flora of the British Isles is very close to 1% of the world's total, producing a total number of wild interspecific vascular plant hybrids in the world of 77 800. The rarity of many hybrids, and the under-studied nature of much of the world's flora, suggests that this figure might be an underestimate rather than an overestimate. This is backed up by the quite spectacular increase (14.5%) in the number of hybrids known in the British Isles between 1975 and 1985. Moreover, by experimentation we know that a great many more interspecific combinations are feasible; in the Orchidaceae alone several tens of thousands of interspecific combinations have been synthesized.

It is, of course, the case that the majority of hybrid combinations are rather rare, but some are frequent and others very common. Two extreme examples, between which every intermediate situation exists, can be taken from the British flora. The hybrid grass *Catapodium marinum* × *C. rigidum* has been found once (in 1960) as a single plant, which was annual and sterile; it has not been found outside Britain and has not been synthesized. On the other hand, in many parts of Britain, over 90% of plants in the *Rosa canina* aggregate are hybrids. In Leicestershire plants of this aggregate are an extremely common feature of virtually every hedgerow and copse, but apparently none is pure *R. canina*. However, abundance of hybrids is not a very good estimate of evolutionary significance. More precisely, it cannot be assumed that rare hybrids are necessarily insignificant, since important amphidiploids such as *Spartina anglica* and *Senecio cambrensis* have arisen from the products of rare hybridization events.

When assessing the abundance of hybrids, as above, one should bear in mind that normally only recent, unequivocally identifiable hybrids are being considered. Much higher figures would be obtained if ancient, stabilized hybrids were taken into account as well.

III. Fertility of Hybrids

As with abundance, every range of fertility can be found in hybrids, from totally sterile to totally fertile. In some well-known hybrids, such as *Geum rivale* × *G. urbanum*, there does not seem to be any reduction in fertility from that shown by either parent. At the other extreme, some hybrids do not produce flowers or, if they do, these are aborted or their stamens or carpels show development arrested prior to spore formation. For example, certain hybrids in *Ranunculus* subgenus *Batrachium* produce flowers without stamens or carpels, while in some *Sorbus* hybrids stamens are produced but the pollen mother cells degenerate into a plasmodial archesporium that does not produce pollen.

Where meiosis does take place, but is irregular, very low levels of fertility are often found — often nil in any given sample. However, it is dangerous to assume nil fertility in such cases. With larger sample sizes some pollen fertility is usually encountered (often a small fraction of 1%), and the same applies to the megaspores. In a triploid plant with $x = 7$ ($2n = 21$), if one assumes that seven trivalents form and that microspores or megaspores with either seven or fourteen chromosomes are viable, then $1/2^6$ (1/64) spores will be functional. In practice, results with be lower than this because it will not be the case that only trivalents will form. But even if they do not there are calculable chances that viable gametes will be formed. Although viable male and female gametes from such a hybrid are very unlikely to meet (perhaps $1/64^2 = 1/4096$ or lower in the above example), the chances of backcrossing can be much higher.

The only conclusion that can be drawn is that, if meiosis does occur (as is the case in the majority of hybrids investigated), there will always be a level of fertility above 0%, even though it might be very low. The *Festuca* hybrid discussed later illustrates this well. In this case a pentaploid ($2n = 35$) hybrid produced triploid ($n = 21$) female gametes in a proportion of at least 1/1500 (Stace and Ainscough, 1984).

Whatever the degree of fertility shown by a hybrid, it must not be assumed that this is a measure of its evolutionary significance. Highly sterile hybrids can often contribute to evolution in a very effective way, especially via introgression, as shown by the *Festuca*. Amphidiploidy is also a very obvious route to evolutionary success via a totally or highly sterile hybrid, as in the *Senecio* and *Spartina* examples mentioned above.

It is also well established that higher levels of fertility can be selected for very rapidly, e.g. in *Gilia* (Grant, 1966a).

IV. Routes of Evolution via Hybridization

Many routes of speciation in particular or of evolution more generally are known to involve hybridization, and have been reviewed by several workers, notably Grant (1953, 1971). Grant envisaged the evolution of new species from a hybrid as a process of stabilization, and for convenience he grouped these processes under seven headings. Different systems of grouping are obtained according to the extent to which various factors are given prominence. For example one could emphasize homoploid versus heteroploid levels; homozygous versus heterozygous states; homogamic versus heterogamic conditions; sexual versus asexual reproduction; or external versus internal isolating mechanisms. These various systems can be used to serve rather different aims.

The aims of the present chapter are not well served by adopting any particular grouping of mechanisms, since it is not intended to review routes of speciation via hybridization. Instead, a few comments will be made as a background to the following section.

Amphidiploidy is by far the best known mechanism. The amphidiploid has the somatic chromosome number of the sum of those of its two parents, having arisen in most cases by the doubling of the chromosome number of the F_1 hybrid. However, amphidiploids can sometimes arise directly by the hybridization of two autopolyploids; the production of a somatic hybrid by protoplast fusion is a closely parallel procedure. In addition, several sorts of derived polyploids are known. For example, in *Centaurium*, a new allohexaploid arose by the backcrossing to one of its parents of a largely infertile tetraploid hybrid between two tetraploid species (Ubsdell, 1979). In many cases derived amphidiploids represent heteroploid introgression. Thus, in *Holcus*, it is likely that the pentaploid representatives of *H. mollis* arose (and are still arising) by hybridization between the tetraploids of that species and the diploid *H. lanatus* (Beddows, 1971).

Amphidiploidy represents one way of overcoming hybrid sterility. Hybrids that have developed successful means of reproduction by vegetative propagation or by apomixis represent others. Like amphidiploids, these sorts of hybrids are well isolated genetically from their parents.

Fertile hybrids, on the other hand, need to develop isolating mechanisms if they are to evolve as entities separate from their progenitors. Examples exist of hybrid taxa (F_1 hybrids, later generation segregants, backcross derivatives, introgressed variants) that have achieved isolation by virtually all of the numerous interspecific isolating mechanisms that are known, including both external and internal and both pre-zygotic and post-zygotic mechanisms.

Finally, mention should be made of two unusual systems. The permanent translocation heterozygotes found in *Oenothera* represent a group of diploid microspecies (perhaps thousands, certainly hundreds), each with a unique combination of two of the large number of known Renner Complexes. They are kept separate by the fact that these plants are mainly inbreeding, and little or no recombination occurs between the two Renner Complexes of a microspecies. Occasional hybridization events generate new Renner Complex combinations.

The permanent odd polyploids of *Rosa* represent pentaploids ($2n = 35$), or less often tetraploids or hexaploids that produce haploid male gametes and tetraploid (less often triploid or pentaploid) female gametes. The three to five extra sets of chromosomes do not normally pair or segregate at meiosis and hence do not show high levels of recombination. It is the different sets of these possessed by the plants that differentiate the microspecies. Hybridization is widespread in this group, and this leads to novel combinations of these non-pairing sets of chromosomes, as can occasional pairing between them.

V. Assessment of Importance of Hybridization in Evolution

The existence of many known routes of evolution involving hybridization throws virtually no light at all on the extent to which these have been exploited, i.e. upon the importance of these routes. There are two different approaches to answering the latter question which may be termed the experimental and the retrospective approaches.

Much can be gained from an experimental attitude of mind, and no self-respecting scientist can work without it, but it cannot provide all the answers. In the case of hybridization, it is never possible to repeat an experiment of nature, since the raw ingredients are now different and conditions have altered. Hence one can never prove that a plant has a particular parentage, nor that plants of similar appearance had the same origin. In my opinion the fullest answers about the origin of hybrids and the role of hybridization in evolution can be obtained only by the dual method of the two above approaches.

A. The experimental approach

An obvious first step is to carry out artificial hybridization between the putative parents of wild hybrids to attempt to resynthesize the latter, and this has proved an enormously successful line of work. Where a hybrid can be recreated in this way the evidence overwhelmingly confirms the hypothesis. The reverse, however, is not true. There are very many cases where extensive attempts to resynthesize hybrids whose parentage is virtually certain have failed, e.g. *Glyceria fluitans* × *plicata, Potentilla erecta* × *reptans*, and *Festuca rubra* × *Vulpia myuros*. The reasons are many. Sometimes the technique has proved too difficult. Often only certain parental genotypes are compatible, and it is a case of finding the right ones to succeed. In some cases the formula for success is quite straightforward, e.g. *Gallium mollugo* and *G. verum* will cross at the tetraploid but not at the diploid level, but finding this is a matter of chance. Frequently plants of hybrid origin do not have a simple parentage (e.g. *Galeopsis tetrahit*) and complicated lineages need to be retraced before resynthesis can be achieved. Finally, many crosses occur with a very low level of success, and it is impossible for man to match the millions of attempts that have taken place in nature.

Hence our experiments at resynthesis provide a rather inaccurate guide to the situation in the wild — although we can produce many hybrids that do not occur in nature, we cannot produce many that do.

When simple crossing experiments are carried further, extra information is often obtained. For example, Grant (1966b) reported that seven of the "numerous" interspecific *Gilia* hybrids he had produced over the years had doubled spontaneously to produce allopolyploid progeny, some repeatedly so, and one was intersterile with both parents.

Long-term studies in the wild can also be very revealing. Heiser (1979) gave successive reports on hybridizing *Helianthus* species in Indiana in 1950, 1961 and 1972, and was able to correlate the differences in the three sites examined with changes in land use. In addition to Heiser's work, and to similar studies mentioned by him in other groups, the development of three very recently formed groups of amphidiploids (*Senecio cambrensis, Spartina anglica* and *Tragopogon mirus* and *miscellus*) have been documented over periods of up to 100 years.

Experimental work has also been very instructive in showing how and to what extent hybrids can exhibit new characters not present in the parents. Good examples are the studies of Alston and Turner (1963) on phenolics in *Baptisia*, Barber (1970) on enzymic proteins in × *Triticosecale*, and Cook (1970) on leaf shape in *Ranunculus* subgenus *Batrachium*. Admittedly one is seeking rather drastic jumps if one is to explain speciation (saltation) in this way, but there is good evidence to indicate that hybridization can act as a trigger for the expression of novel characters.

B. *The retrospective approach*

The experimental approach can demonstrate the reality of the different evolutionary routes via hybridization, i.e. show us their qualitative importance, and indicate the expected outcome of the different sorts of hybridization events. From this information the retrospective investigation of wild populations should indicate the quantitative importance of the various routes.

Of all the known routes, amphidiploidy is the easiest to quantify; the proportion of amphidiploids among wild plants today must be a direct measurement of the frequency with which that route has been exploited. Estimates of the proportion of polyploids among angiosperms vary from about one-third to three-quarters, most of this variation being caused by the interpretation of palaeopolyploidy. Neopolyploids, i.e. plants that are obviously at a polyploid level because they have known diploid relatives, probably amount to 30–35% of all flowering plants (Stebbins, 1950). The total percentage is increased above this level according to how many plants are considered to be palaeopolyploids.

The genus *Petasites* (Asteraceae), for example, consists of 18 species, all of which have the chromosome number $2n = 58$ ("diploids") or 87 ("triploids"), i.e. $x = 29$ (Toman, 1972). Chromosome numbers in genera closely related to *Petasites*, i.e. *Endocellion, Tussilago* and *Homogyne*, are equally high or higher (up to $2n = 160$). Obviously all of these plants are palaeopolyploids, and should be counted as polyploids in any survey. Many other plants fall in the same category, e.g. *Salix* $2n \geq 38$ (500 species), *Magnolia* $2n \geq 38$ (80 species). However, there is debate as to the highest level that can be considered diploid. One suggestion (Grant, 1963) is that plants with $n > 13$ should be considered diploid in origin, and this seems to me a reasonable number. On this basis Grant calculated that 43% of dicotyledons and 58% of monocotyledons (47% of angiosperms overall) are polyploid. Since 1963 the number of species counted has at least doubled, and Goldblatt (1980) recalculated the percentage of polyploid monocotyledons to be 55%. Using a lower permissable level for diploidy ($n < 11$) Goldblatt reckoned that 68% of monocotyledons were polyploid, and Lewis (1980) estimated a similar figure for dicotyledons. Hence, on conservative estimates about half of all angiosperms are polyploid in origin, and the figure could be as high as 70%.

These are raw data, but before firm deductions can be made further questions need to be answered. It is not intended to answer all these here, but the more important questions must be formulated, in order to gain an idea of the sorts of problems that exist.

How many polyploids arose from diploids, as opposed to arising from other polyploids? Even if most arose along the latter path, polyploidy could not be said to be any less important in evolution, but the pattern of speciation would be rather different in the two situations. It is possible that cladistic methods might be able to give an answer to this problem by elucidating the phylogenetic pattern in groups containing diploids, neopolyploids and palaeopolyploids.

How true is it that a bivalent-forming polyploid (i.e. an amphidiploid or segmental allopolyploid with multivalent suppressors) is of hybrid origin? To answer this question we need a better method of genome analysis than is available at present. Possibly the approach advocated by Kimber (1984) will assist here. Stebbins (1985) believes that most polyploids are interspecific or interracial hybrids, because by experiment one can show that autopolyploids are generally not good competitors.

How can we answer the conundrum posed by the facts that at least 50% of plants are polyploid, yet diploids still occur in all of the major angiosperm groups (Lewis, 1980)? If one discounts a reverse evolution from polyploidy to diploidy, one could argue against the commonly stated tenet that evolution proceeds today in much the same way (i.e. using the same mechanisms) as it did tens or hundreds of millions of years ago, because it seems possible that early evolution of the main angiosperm groups was at the diploid level, and at a later time polyploidy has become a commoner route. Grant (1953) suggested that homoploid complexes were the important ones in evolution, and that polyploidy has been superimposed. Yet many primitive groups are wholly or mainly palaeopolyploid, e.g. Magnoliales, and Poaceae-Bambusoideae (Stebbins, 1985). An attractive answer to this anomaly is partially provided by the idea of the hybridization cycle (Ehrendorfer, 1959; Stebbins, 1969; Carson, 1985), with different groups of plants at different stages in it.

Since certainly 50% and perhaps 70% of all angiosperms are polyploid and almost certainly arose by hybridization, the latter process must have been one of the main routes in evolution. If we add to the evidence from polyploidy that from all the other routes of speciation listed in the previous section, but which are far more difficult to quantify, then it becomes clear that plant evolution has a very highly reticulate pattern. The realization of this fact is posing major problems in the cladistic analysis of higher plants that are still to be resolved (Funk, 1985).

VI. Instructive Examples of Speciation

Two examples will be discussed briefly here. They both throw a certain amount of light on the present topic, but in each case there are areas of uncertainty and each demonstrates the often unexpected and complicated relationship between hybridization and evolution.

A. Senecio squalidus *and* S. vulgaris

The occurrence of sterile triploid hybrids between these two species, and their spontaneous chromosome doubling to form the new hexaploid amphidiploid *S. cambrensis* (Rosser, 1955; Weir and Ingram, 1980; Abbott *et al.*, 1983; Ingram and Noltie, 1984) (Fig. 1) are well documented. The relationship between the

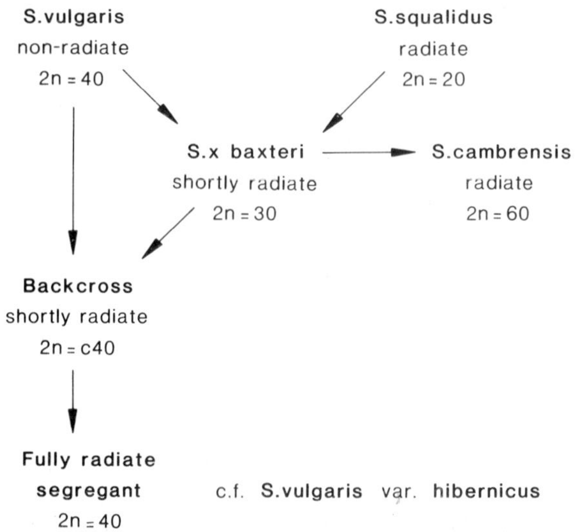

FIG. 1 Probable origin of *Senecio cambrensis* and possible origin of *S. vulgaris* var. *hibernicus*.

two parental species that I shall discuss here is not concerned with *S. cambrensis*, but with a remarkable phenomenon that has been witnessed in Britain over the past 20 years. In that period the radiate variant of *S. vulgaris*, previously rare and sporadic, has become a common and conspicuous feature of waste ground in many areas. The radiate variant is, like the non-radiate plant, tetraploid, and the two morphs are completely interfertile. The radiant gene is incompletely dominant, hybrids possessing much reduced ray florets and segregating progeny in the expected 1:2:1 ratio.

There are two possibilities for the origin of radiate plants: mutation of non-radiate plants; and introgression from another, radiate species, presumably the diploid *S. squalidus* (Fig. 1). The evidence at least up to 1979, was equivocal (Stace, 1977). However, Ingram *et al.* (1980) reported that they had been able to synthesize plants "resembling" radiate *S. vulgaris* from an artificial *S. vulgaris* × *S. squalidus* by backcrossing this to *S. vulgaris* and then selfing the products (Fig. 1). Progeny were grown from two backcross derivatives; from one no truly radiate plants were obtained, from the other 7 out of 27 plants had normal ray-florets. Four of these seven plants were examined cytologically: one had $2n = 60$, one had $2n = 48$, and two were tetraploid or near tetraploid ($2n = 40$ to 42), with regular meiosis. Fertility of these plants ranged from 10% to 70% (mean 30%).

The careful experiments of Ingram *et al.* have shown clearly that the origin of radiate *S. vulgaris* via hybridization with *S. squalidus* is very plausible. The origin of a tetraploid radiate *S. vulgaris* from a triploid hybrid in only two generations is very remarkable. The synthetic radiate plant came "within the range of variation attributed to *S. vulgaris*", suggesting that it was indistinguishable from wild radiate *S. vulgaris*, which has not been shown to differ from non-radiate plants in other characters. However, the wild and synthetic plants are not identical, since the fertility of the latter was only up to 70%, while that of the former is 100%.

This remarkable story illustrates one important general point. Here is a case of a well-documented evolutionary event taking place at the present time ("evolution in action"), backed up by well-organized experiments, yet still we cannot be quite certain whether radiate *S. vulgaris* has arisen via hybridization, mutation, or both. Further experiments are needed, particularly using wild-collected seed from natural hybrids rather than using the products of artificial crosses. But no amount of evidence can *prove* the exact course of a past event, even when the raw materials are still available to us. In the majority of cases the raw materials are no longer available, and we are still further from certainty.

B. Festuca rubra *and* Vulpia fasciculata

The genus *Festuca* is large and diverse, and has been divided into several sections by various authors, although a modern overall synthesis is lacking. Experiments have shown that hybridization is possible between some of these sections, and between *Festuca* and two other genera, *Lolium* and *Vulpia*. Moreover, hybrids between several of these taxa are known in the wild (Fig. 2). For present purposes I shall discuss only the hybrid between the hexaploid *Festuca rubra* (*Festuca* section *Festuca*) and the tetraploid *Vulpia fasciculata* (*Vulpia* section *Monachne*). This pentaploid hybrid is frequent on coastal dunes in Britain, and easily synthesized with the *Vulpia* as the female parent. I shall refer to it as the pentaploid *Festulpia*.

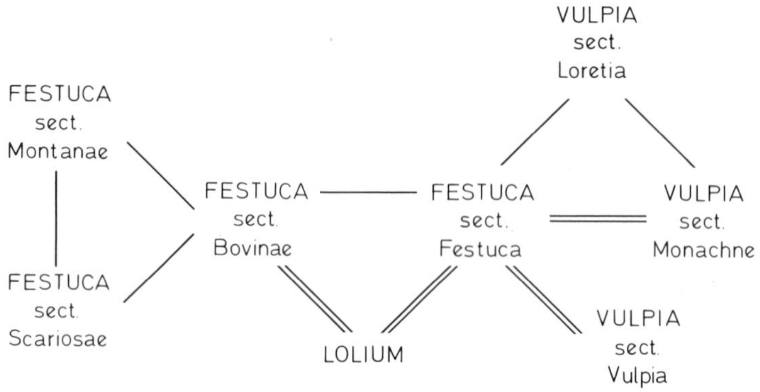

FIG. 2 Intersectional and intergeneric hybrid combinations in the genera *Festuca, Lolium* and *Vulpia*. Single line, artificial only; double line, natural and artificial.

Like most of the above intersectional and intergeneric hybrids, the pentaploid *Festulpia* is highly sterile (pollen stainability $< 1\%$). It is not, however, completely sterile. From an open pollinated plant in cultivation approximately 1500 ripened florets were sown, and these produced one mature plant (Stace and Ainscough, 1984). This was a hexaploid and must have been a backcross to *F. rubra*, pollen of which was available to the pentaploid *Festulpia* in great abundance. The most likely route of production of this backcross derivative is shown in Fig. 3. In that scheme it is assumed that chromosome pairing in the pentaploid was homogenetic (F−F and V−V). The backcross derivative is fertile and presents highly regular bivalent formation, which, under the above assumption, must include heterogenetic pairing (F−V). Alternatively, heterogenetic pairing might have occurred in the pentaploid.

In either case this is an example of heterogenetic pairing occurring between the genomes of different genera. A new hexaploid that is fertile and closely similar to some variants of *F. rubra* has evolved by introgression from a tetraploid *Vulpia* into parental hexaploid *F. rubra* via a highly sterile F_1 hybrid. It is particularly interesting because the *F. rubra* aggregate is a declining polyploid complex, and over most of their area of distribution only hexaploids and octoploids exist. Wide hybridization has in this case permitted continued enrichment of the gene pool of *F. rubra* which would not be possible if only species of *Festuca* were involved.

I believe the important lesson in this example to be that hybridization can occur between only distantly related taxa, yet produce evolutionally important results.

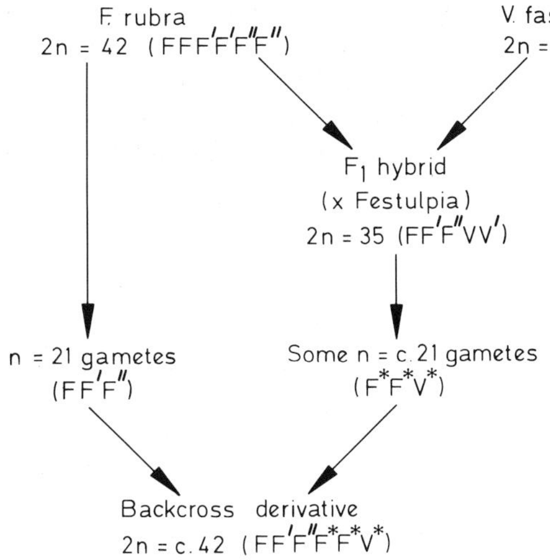

FIG. 3 Hypothetical scheme for the production of a hexaploid backcross derivative of a pentaploid × *Festulpia* hybrid, assuming homogenetic pairing in the pentaploid. F, F′ and F″, and V and V′, represent homoeologous *Festuca* and *Vulpia* genomes. F* and V* represent mixed *Festuca* and *Vulpia* genomes produced by independent assortment and crossing over.

Hence, in measuring the importance of hybridization in evolution, we must look well beyond the limits that on first impressions appear to be relevant.

The extent to which the *Festulpia* situation is found in nature is not easy to establish, but other examples are known, both in the wild and in plant breeding studies, and there is no reason to believe that it is not a widespread phenomenon. In *Triticale* breeding, for example, the most promising lines today are secondary hexaploids. These are mostly produced by backcrossing to a primary hexaploid the heptaploid F_1 hybrid between a primary hexaploid and a primary octoploid (Fig. 4).

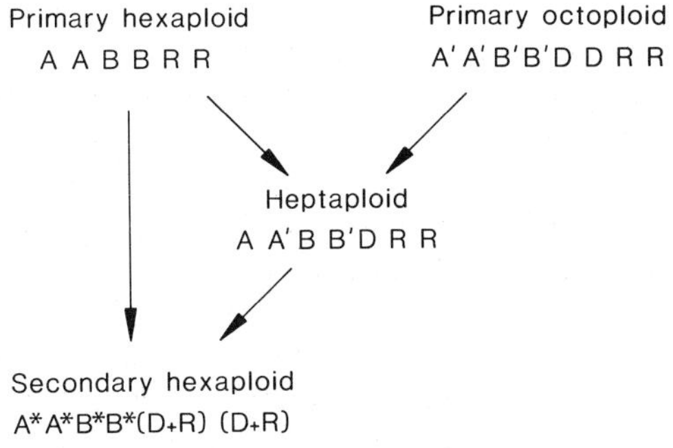

FIG. 4 Scheme for the production of secondary hexaploid *Triticale*. A and B are genomes from tetraploid *Triticum*, R those from *Secale*, and A′, B′ and D those from hexaploid *Triticum*. A* and B* represent genomes produced by independent assortment between A and A′, and B and B′, genomes. (D + R) represents a mixed genome produced by substitution of D and R chromosomes.

The formation of a new hexaploid from the cross between a heptaploid and a hexaploid is closely analagous to the production of a hexaploid *Festuca* from a hexaploid × pentaploid cross. In other respects there are differences, notably in relation to the level of genome homology. There is no homology between the D and R genomes found in heptaploid and octoploid *Triticale*, so in the secondary hexaploids substitution lines are formed, rather than the truly mixed segregating genomes found in the secondary hexaploid *Festuca*. It seems likely, however, that substitution lines also occur in nature, and might in fact play a major part in genome evolution. This would explain, for example, the widely different levels of homology often found between the different homoeologues of two genomes. *Triticale* "Armadillo" arose from a spontaneous crossing of hexaploid *Triticale* with hexaploid *Triticum*, and possesses a 2R/2D substitution, showing that substitution lines can originate without man's manipulation.

VII. Conclusions

Hybrids are common in nature and all the evidence we have indicates that hybridization has been one of the major factors in angiosperm evolution. At least 50% of all angiosperm species have evolved via a route involving hybridization. Despite the unequivocality of the general situation, it is rarely possible to be certain about the courses of evolution in particular cases. Numerous and diverse experiments have told us a great deal, but hybridization has often taken place in a complicated and unpredicable way, and extremely extensive studies are usually required before even an outline of the probable routes can be obtained. Further detailed experiments and observations are required over a wide range of plant groups in different regions of the world, especially the tropics.

References

Abbott, R. J., Ingram, R. and Noltie, H. J. (1983). Discovery of *Senecio cambrensis* Rosser in Edinburgh. *Watsonia* **14**, 407–408.

Alston, R. E. and Turner, B. L. (1963). Natural hybridization among four species of *Baptisia* (Leguminosae). *Am. J. Bot.* **50**, 159–173.

Barber, H. N. (1970). Hybridization and the evolution of plants. *Taxon* **19**, 154–160.

Beddows, A. R. (1971). The inter- and intra-specific relationships of *Holcus lanatus* L. and *H. mollis* L. *sensu lato* (Gramineae). *Bot. J. Linn. Soc.* **64**, 183–198.

Carson, H. L. (1985). Unification of speciation theory in plants and animals. *Syst. Bot.* **10**, 380–390.

Cook, C. D. K. (1970). Hybridization in the evolution of *Batrachium*. *Taxon* **19**, 161–166.

Ehrendorfer, F. (1959). Differentiation-hybridization cycles and polyploidy in *Achillea*. *Cold Spring Harbor Symp. Quant. Biol.* **24**, 141–152.

Funk, V. A. (1985). Phylogenetic patterns and hybridization. *Ann. Mo. Bot. Gard.* **72**, 681–715.

Goldblatt, P. (1980). Polyploidy in angiosperms: monocotyledons. *In* "Polyploidy: Biological Relevance" (W. H. Lewis, ed.), pp. 219–239. Plenum Press, New York.

Grant, V. (1953). The role of hybridization in the evolution of the leafy-stemmed Gilias. *Evolution* **7**, 51–64.
Grant, V. (1963). "The Origin of Adaptations". Columbia University Press, New York and London.
Grant, V. (1966a). Selection for vigor and fertility in the progeny of a highly sterile species hybrid in *Gilia*. *Genetics* **53**, 757–776.
Grant, V. (1966b). The origin of a new species of *Gilia* in a hybridization experiment. *Genetics* **54**, 1189–1199.
Grant, V. (1971). "Plant Speciation". Columbia University Press, New York and London.
Heiser, C. B. (1979). Hybrid populations of *Helianthus divaricatus* and *H. microcephalus* after 22 years. *Taxon* **28**, 71–75.
Ingram, R. and Noltie, H. J. (1984). Ray floret morphology and the origin of variability in *Senecio cambrensis* Rosser; a recently established allopolyploid species. *New Phytol.* **96**, 601–607.
Ingram, R., Weir, J. and Abbott, R. J. (1980). New evidence concerning the origin of inland radiate groundsel, *S. vulgaris* L. var. *hibernicus* Syme. *New Phytol.* **84**, 543–546.
Kimber, G. (1984). Evolutionary relationships and their influence on plant breeding. *In* "Gene Manipulation in Plant Improvement", 16th Stadler Genetics Symposium (J. P. Gustafson, ed.), pp. 281–293. Plenum Press, New York and London.
Lewis, W. H. (1980). Polyploidy in angiosperms: dicotyledons. *In* "Polyploidy: Biological Relevance" (W. H. Lewis, ed.), pp. 241–268. Plenum Press, New York and London.
Rosser, E. M. (1955). A new British species of *Senecio. Watsonia* **3**, 228–232.
Stace, C. A. (1975a). "Hybridization and the Flora of the British Isles". Academic Press, London and New York.
Stace, C. A. (1975b). Wild hybrids in the British flora. *In* "European Floristic and Taxonomic Studies" (S. M. Walters, ed.), pp. 111–125. Botanical Society of the British Isles, London.
Stace, C. A. (1977). The origin of radiate *Senecio vulgaris* L. *Heredity* **39**, 383–388.
Stace, C. A. and Ainscough, M. M. (1984). Continuing addition to the gene-pool of the *Festuca rubra* aggregate (Poaceae:Poeae). *Plant Syst. Evol.* **147**, 227–236.
Stebbins, G. L. (1950). "Variation and Evolution in Plants". Columbia University Press, New York and London.
Stebbins, G. L. (1969). The significance of hybridization for plant taxonomy and evolution. *Taxon* **18**, 26–35.
Stebbins, G. L. (1985). Polyploidy, hybridization and the invasion of new habitats. *Ann. Mo. Bot. Gard.* **72**, 824–832.
Toman, J. (1972). A taxonomic survey of the genera *Petasites* and *Endocellion. Folia Geobot. Phytotax., Praha* **7**, 381–406.
Ubsdell, R. A. E. (1979). Studies on variation and evolution in *Centaurium erythraea* Rafn and *C. littorale* (D. Turner) Gilmour in the British Isles, 3. *Watsonia* **12**, 225–232.
Weir, J. and Ingram, R. (1980). Ray morphology and cytological investigations of *Senecio cambrensis* Rosser. *New Phytol.* **86**, 237–241.

Part III

DIFFERENTIATION IN DIVERSE BIOLOGICAL SYSTEMS

7 Genetic Structure and Diversity Patterns in Adjacent Populations: *Arrhenatherum elatius* (L.) Beauv.

PHILIPPE VERNET, ALEXIS DUCOUSSO, DANIEL PETIT and MYRIAM VALERO

Université des Sciences et Techniques de Lille, Laboratoire de Génétique Ecologique et de Biologie des Populations Végétales, Villeneuve d'Ascq, France

I. Introduction

Selective forces, neutral events, and breeding system combined with population size and history of taxa, represent essential factors modulating genetic diversity. The idea that high levels of genetic diversity are positively related with a spatio-temporally heterogenous and uncertain environment is well supported by comparative data on both plant and animal species (Nevo, 1978; Maynard-Smith, 1978; Hamrick *et al.*, 1979). It seems that a frequent, perhaps the most common, reaction to disturbance in the natural genetic system is to increase the diversity in different directions, especially by modifying the breeding system. In *Thymus vulgaris*, for example, the highest proportions of male-sterile individuals (obligate outcrossing plants) are found in disturbed environments (Valdeyron *et al.*, 1973; Gouyon and Vernet, 1982; Gouyon *et al.*, 1983). In *Dactylis glomerata*, a population recently introduced by migration into a plain of the French mediterranean region increases its diversity by incorporating genes from the neighbouring native subspecies into its gene pool (Lumaret and Valdeyron, 1980; Gouyon *et al.*, 1983; Valero and Olivieri, 1985). Nevertheless, populations composed of few individuals, particularly those resulting from a bottleneck in the recent past, have been shown to contain very little diversity (Bazzaz, 1983; Ledig and Conckle, 1983).

Differentiation Patterns in Higher Plants
ISBN: 0-12-709425-3

In the absence of selection, restriction of gene flow can cause random genetic drift leading to genetic differentiation (Ennos, 1984; Kimura and Maruyama, 1984). Extended gene flow is expected to have the opposite effect, namely limiting both inbreeding and differentiation. Species and local populations with very open breeding systems associated with wide dispersal of pollen and seeds may have more genetic variability than selfers (Solbrig, 1972; Jain, 1976). The impact of artificial selection on genetic diversity is largely unknown (Brown, 1979). It seems, however, that allozyme variation in cultivars may either be equal to that in wild populations (Levin, 1976, 1978) or less pronounced (Ellstrand and Marshall, 1985).

The extreme habitats created by man as a result of mining and other industrial activities seem to be inhabited by species which are very variable genetically, 90% of the variation being detected within populations (e.g. *Silene maritima* and *S. cucubalus*: Verleij *et al.*, 1985).

As far as the evolution of adjacent populations is concerned, tolerant populations of *Agrostis tenuis* and *Anthoxanthum odoratum* which have successfully colonized heavy metal-contaminated sites are genetically different from adjacent populations of normal habitats, though these two types of populations are connected by gene flow. Two mechanisms which reduce gene flow are phenological shift (McNeilly and Antonovics, 1968) and an increase of self-fertilization (Antonovics, 1968). It seems that a possible linkage between selected genes and reproductive isolation genes might lead to genetic divergence (Caisse and Antonovics, 1978). We should expect tolerant populations to be far less polymorphic than "normal" adjacent populations, (i) because of more or less unique founder effects (small number of individuals at the birth of population) followed by strong selective pressure, and (ii) because the populations of toxic habitats are protected against gene flow. In this sense, results obtained by Wu *et al.* (1975), who found in *Agrostis stolonifera* a higher phenotypic diversity (morphological and isozymatic) in tolerant populations than in normal ones, raise a paradox. This is only partly explained by the suggestion that tolerant types must not be very rare in normal pasture populations from which tolerant populations have evolved (Wu *et al.*, 1975), because the genetic population diversity is even higher in toxic habitats than in normal pastures. The tolerant populations should reflect the high diversity of disturbed habitats.

The aim of the study presented in this chapter was to discover if the paradox with respect to diversity exists in *Arrhenatherum elatius* growing on adjacent normal and toxic soils in northern France.

II. The Species and Sites Studied

Arrhenatherum elatius is a perennial stoloniferous grass. This autotetraploid ($4x = 28$), wind pollinated and outcrossing species is able to colonize a wide range of habitats in the north of France (fallow land, meadow, dune, spoil heap, calaminary waste land, etc.).

Morphological and electrophoretic (allozyme and isozyme) characteristics were studied in three sites where adjacent tolerant (T) and non-tolerant populations (N) are in contact. These sites are located in the mining coal field basin of northern France (Fig. 1). Site 1 represents a calaminary waste land characterized by high heavy metal content (lead, zinc, cadmium, etc.) of the soil. Sites 2 and 3 are situated on colliery spoil heaps characterized by a disequilibrium in the magnesium/calcium ratio (Mg/Ca > 0.88).

To compare mine populations and pasture populations, transects from toxic mine waste (T) to normal habitats (N) were established (Fig. 1). The characteristics of the sites studied are given in Table 1. Total content in zinc and lead, measured in the soil between the roots of *Arrhenatherum elatius*, varied greatly along the transects (Fig. 2).

III. Level of Diversity in Pasture and Mine Populations

Our study deals, on the one hand, with variation in some physiological and morphological characters and, on the other hand, with genetic variation in different enzyme systems.

A. Variation in physiological and morphological traits

1. Phenology. Plants can be characterized by the timing of the flowering period and by the duration of this period (Fig. 3); the phenological diversity is clearly higher on the toxic soils. Another estimation of the phenological diversity has been made by sampling populations at the same period and by sorting tillers between 10 phenological classes (Table 2). From these data (see Table 3) it appears that the number of classes is lower in pasture soils than in mine soils. The Shannon index, described in detail on p. 139, quantifies this difference (Table 3).

2. Colour and hairiness of panicle node. Table 4 summarizes the distribution of node phenotypes between populations. Individuals with coloured nodes are more abundant on mine than on pasture soils. The "normal" populations look much more uniform than those occurring on mine waste. By contrast, differences between adjacent populations in regard to hairiness are negligible. However, an increase in the number of individuals with hairy nodes can be noted at site 3T. The I_s index (Table 3) underlines the higher diversity of mine soil populations.

3. Plant height. In the pasture habitats tall individuals are the most common; the mean is very high and varies from 134 − 142 cm (Table 5). In parallel, size diversity is very low (Table 3). By contrast, in the toxic habitats the mean size is relatively low (50 − 92 cm) but the variance is large. The I_s index is again higher in the toxic habitats than in the normal ones.

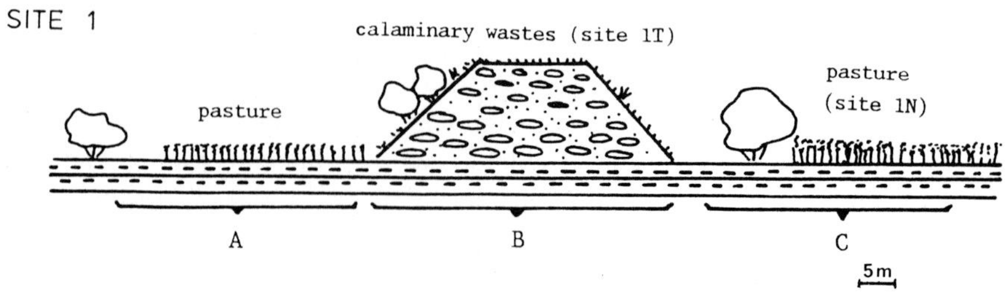

The Auby transect : one plant studied each metre in A, B and C.

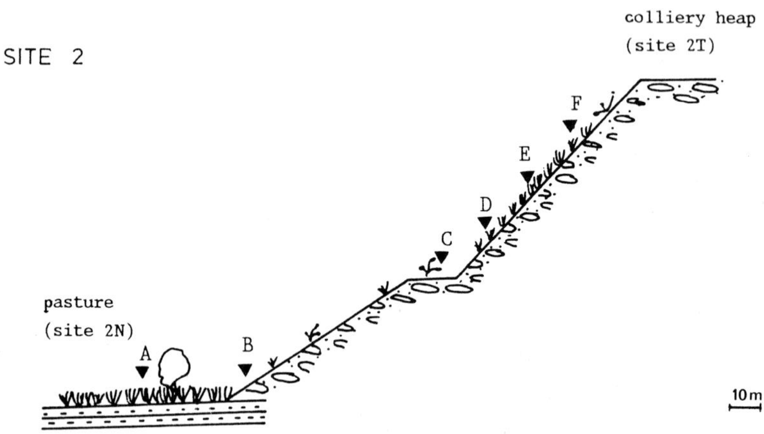

The Hénin-Beaumont transect . A, B, C, D, E, F : one plant studied each metre, over about 20 metres.

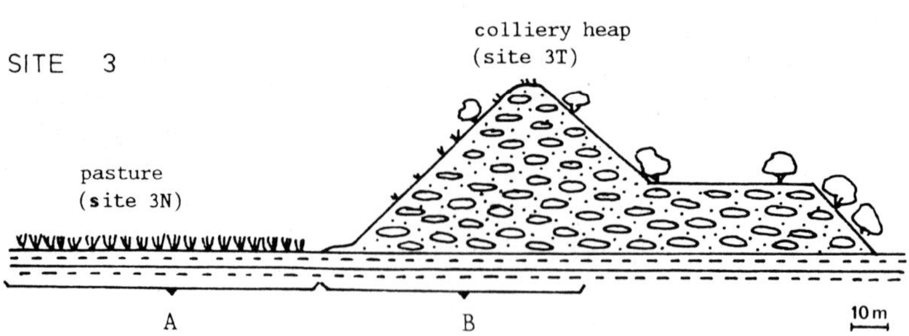

The Auchy transect : one plant studied each metre in A and B .

FIG. 1 Transects across the boundaries between mine and pasture showing the sites sampled. T, toxic soil; N, normal soil.

TABLE 1
Main characteristics of the adjacent habitats studied

Code	Soil type	Heavy metal content (p.p.m.)		Mg/Ca ratio	Density of *Arrhenatherum* populations (panicles m^2)	Number of species	Dominant species
		Zn	Pb				
1	Normal (N)	1919	75	0.14	61	22	*Arrhenatherum elatius*
	Mine (T) (Pb, Zn)	5865	463	0.095	15	6	*Armeria maritima* *Arrhenatherum elatius*
2	Normal (N)	4.8	4.6	0.17	352	21	*Arrhenatherum elatius*
	Mine (T) (schist)	—	5.2	0.88	69	11	*Rumex scutatus* *Arrhenatherum elatius*
3	Normal (N)	—	1	0.089	85	22	*Dactylis glomerata*
	Mine (T) (schist)	—	0.7	2.1	1.6	13	*Pastinaca sativa* *Arrhenatherum elatius*

N, normal; T, toxic.

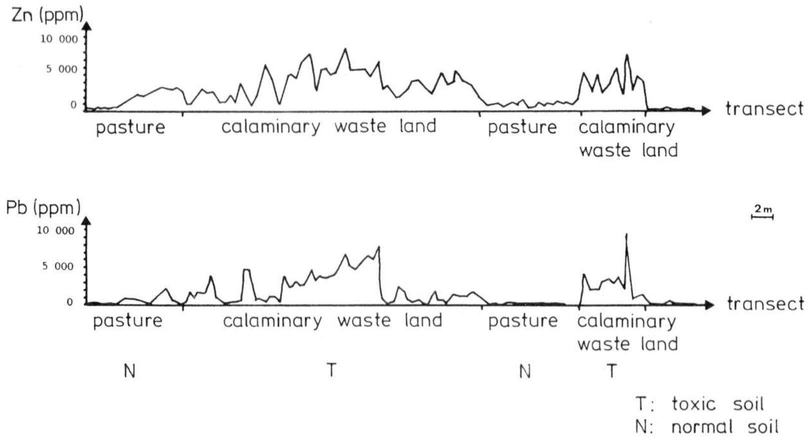

FIG. 2 Variability in lead and zinc content along the Auby transect. Soil samples taken every metre.

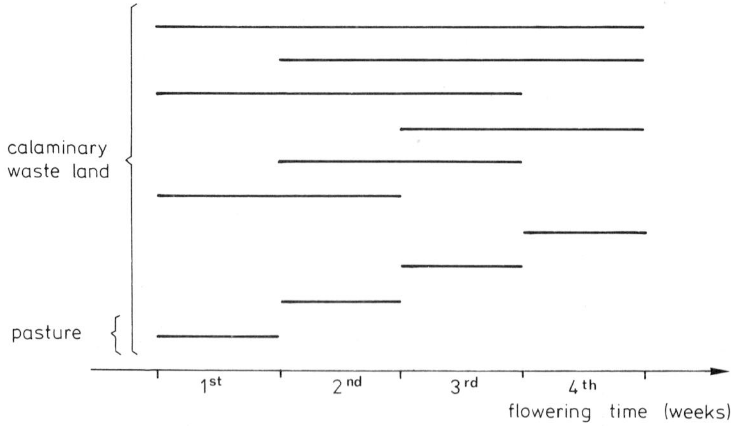

FIG. 3 Calaminary waste land and pasture. Flowering time of plants representing ten different phenological types.

TABLE 2

Distribution frequencies of individuals among flowering time classes in the three pairs of adjacent populations (June 1983)

Soil type	Site code	Number of individuals sampled	Number of tillers sampled	Phenological classes									
				1	2	3	4	5	6	7	8	9	10
Pasture	1N	109	2490			0.02		0.02		0.10	0.67	0.11	0.08
	2N	193	3505			0.04	0.01	0.01		0.02	0.53	0.17	0.22
	3N	211	4406							0.06	0.41	0.33	0.20
Mine	1T	355	1476	0.6	0.04	0.09	0.01	0.01	0.01	0.06	0.06	0.04	0.08
	2T	812	4605	0.07	0.01	0.09	0.01	0.01	0.01	0.06	0.62	0.10	0.02
	3T	555	3347	0.51	0.02	0.07	0.02	0.01	0.01	0.07	0.19	0.07	0.03

1: tillering
2: one node visible
3: two nodes visible
4: appearing of the last leaf
5: ligule visible
6: sheath opened
7: panicle emerged
8: panicle free
9: beginning of the flowering
10: end of the flowering

TABLE 3
Number of classes or phenotypes and diversity index (I_S) for phenology, node character of floral spike, and plant height in pasture and mine populations.

Soil type	Site code	Phenology: Number of classes observed	Phenology: I_s	Node character: Number of phenotypes observed	Node character: I_s	Plant height: Number of classes observed	Plant height: I_s
Pasture	1N	6	1.483	16	2.101	6	1.997
	2N	7	1.727	18	2.310	5	1.341
	3N	4	1.756	15	1.775	6	1.863
Mine	1T	10	2.152	54	4.130	14	3.167
	2T	10	1.964	36	3.909	8	2.459
	3T	10	2.281	47	3.828	8	2.501

Differences in plant height, colour and hairiness of floral spike nodes and phenology in adjacent habitats were maintained under uniform conditions when plants were grown from tillers (Ducousso, 1985, data on the colour node character obtained in sunny conditions), and from seeds (Loufti, unpub. data on the colour character and plant height). This behaviour strongly suggests a genetic component to the control of these variable traits, though precise bases, probably polygenic, are as yet unknown.

B. *Isozyme and allozyme variation*

To obtain an estimate of the level of genetic variation, different enzyme systems were examined using vertical polyacrylamide gel electrophoresis. The electrophoretic study was performed according to Orstein (1964) and Davies (1964), as modified by Gasquez and Compoint (1976). Some of the enzyme systems had to be discarded because unclear or irreproductible enzyme banding patterns appeared.

The genetic basis for the gel banding patterns has been determined by comparisons of isozyme phenotypes from selfing or outcrossed progenies with those of their parents (Ducousso, 1985; Ducousso and Loufti, unpub.). The allozyme polymorphism was found for three loci, namely GOT 1 (with five alleles: 1.08; 1.04; 1.00; 0.92; 0.84 coding for a dimeric enzyme), GOT 2 (with seven alleles: 1.05; 1.00; 0.94; 0.88; 0.82; 0.80; 0.70 coding for a dimeric enzyme), and ADH (with three alleles: 1.12; 1.00; 0.88 coding for a monomeric enzyme). As to the TO system, for which the genetic basis of the banding patterns is still unknown, only the isozyme phenotypes have been used for estimating diversity. The following results were obtained:

1. The TO system. Only three to four phenotypes are found in pasture soils whereas seven to 11 are revealed on mine soils (Table 6). A clear-cut higher diversity of mine soil populations is observed. Plants have not been scored for TO at site 3.

2. Number of alleles, and allelic frequencies at ADH, GOT 1 and GOT 2 loci. The level of genetic diversity of the polymorphic enzyme loci is higher on mine soils for both ADH (Table 7) and GOT 2 (Table 9). At the ADH locus, only one allele (1.00) was frequent at more than 81% on pasture soils, whereas on mine soils its frequency was below 60%, two other alleles (1.12 and 0.88) also being frequent. For the GOT 2 locus, the higher diversity on mine soils was due to infrequent alleles (two to four), absent or exceedingly rare on pasture soils. The I_s index is not sensitive enough to weight this diversity difference. The only exception is that locus GOT 1 (Table 8) is apparently less polymorphic on toxic soils. This situation does not seem to be related to a selection effect. Indeed, none of these loci seem to be undergoing selection effects in the studied sites (Ducousso, 1985). The differential diversity with ADH and GOT 2 loci is repeated at the genotypic level (for further details see Ducousso *et al.*, submitted).

IV. Diversity and Genotypic Structure Analysis

A. Assessment

In order to estimate diversity in the different populations, Shannon and Weaver's entropy index (1963) was used for the various characters, either on phenotypes or genotypes, or on allelic frequencies. This index I_s is given by the formula:

$$I_s = \sum_{i=1}^{n} -x_i \log_2 x_i$$

where x_i is the frequency of the ith class of the characters studied in the population and n the number of classes.

Wright's inbreeding coefficient, F, modified by Lumaret (1981) for applying to tetraploid taxa, has been used to estimate deviations from Hardy–Weinberg expectations:

$$F = 1 - \frac{H_o}{H_e}$$

where H_e and H_o are respectively the expected and observed proportions of heterozygotes.

In the tetraploid *Arrhenatherum elatius*, H_e can be estimated by the proportion of heterozygote gametes, that is

TABLE 4
Distribution of the node phenotypes in the three pairs of adjacent populations

Site 1

Normal soil (109 plants observed)

	ggg	gCC	gCg	ggC	CCg	Cgg	gCC	CCC
sss	37	39			9	10		5
Hss	1	1						
sHs								
ssH		1						
HHs								
HsH	3	5			2	1		1
sHH								
HHH	3	1			1			

Mine soil (355 plants observed)

	ggg	gCC	gCg	ggC	CCg	Cgg	gCC	CCC
sss	2	85	8	1	70	5	3	75
Hss	3	10	5	1	14	2	2	18
sHs	2	5	3	1	4	3	2	5
ssH		1			2			3
HHs	3	9	4	1	7	5	3	9
HsH	1	1	1		1	1	1	
sHH		1	1		1			2
HHH	1	3	1	1	2	1	1	3

Site 2

Normal soil (193 plants observed)

	ggg	gCC	gCg	ggC	CCg	Cgg	gCC	CCC
sss	69	1	45		1			
Hss	8		5					
sHs								
ssH	2							
HHs	2		1		1			
HsH	3		2					
sHH	3		1					
HHH	4	1	3		1			

Mine soil (812 plants observed)

	ggg	gCC	gCg	ggC	CCg	Cgg	gCC	CCC
sss	110	18	141	3	26	3	3	12
Hss	20	4	29		12	1		3
sHs	2		3		2			
ssH	11	4	17		11		2	3
HHs	5		9		2			
HsH	6		13		4			
sHH	4		1					
HHH	21		25		4	1		2

Site 3

	Normal soil (211 plants observed)									Mine soil (555 plants observed)							
	ggg	gCC	gCg	ggC	CCg	Cgg	gCC	CCC		ggg	gCC	gCg	ggC	CCg	Cgg	gCC	CCC
sss	152		67	28				6	sss	15	7	64		191	3	7	94
Hss	4								Hss	3	4	2		8	1	3	4
sHs	2								sHs								
ssH	2								ssH	1	3	2		5	5	1	4
HHs									HHs	3	1	3		6	2	1	4
HsH	5		2	1				1	HsH	2	2	9		16	1	2	7
sHH									sHH	1	1	1		2			1
HHH	6		3	2				1	HHH	2	1	6		12	3	1	6

Both colour phenotypes (g, green; C, coloured) and hairiness phenotypes (s, glabrous; H, hairy) are described by three letters corresponding to colour and hairiness at three levels: first letter is the phenotype at the bottom of the node; second letter is the node phenotype; third letter is the phenotype at the top of the node.

TABLE 5

Distribution frequencies of individuals within height plant classes in three pairs of adjacent populations

Soil type	Site code	N	Plant height classes (cm)															
			11–20	21–30	31–40	41–50	51–60	61–70	71–80	81–90	91–100	101–110	111–120	121–130	131–140	141–150	151–160	161–170
Pasture	1N	442											0.02	0.18	0.47	0.20	0.11	0.02
	2N	524												0.13	0.65	0.21	0.01	
	3N	385										0.01	0.13	0.16	0.51	0.18	0.01	
Mine	1T	191		0.005	0.02	0.025	0.04	0.06	0.10	0.19	0.22	0.15	0.10	0.06	0.02	0.01		
	2T	379				0.01	0.03	0.05	0.17	0.27	0.24	0.21	0.20					
	3T	280	0.01	0.05	0.12	0.29	0.27	0.17	0.07	0.02								

TABLE 6
TO system: number and frequency of phenotypes, diversity index (I_s) in pasture and mine populations

Soil type	Site code	Number of individuals sampled	1.23	1.11	1.00	0.92	1.23 1.11	1.23 1.00	1.23 0.92	1.11 1.00	1.11 0.92	1.00 0.92	1.23 1.11	1.23 1.11 0.92	1.23 1.00 0.92	1.11 1.00 0.92	1.23 1.11 1.00 0.92	Number of phenotypes	I_s[a]
Pasture	1N	35	0.03		0.77					0.11		0.09						4	1.105
	2N	35			0.82							0.09	0.09					3	0.861
Mine	1T	37		0.27	0.16		0.19			0.24			0.08		0.03	0.03		7	2.478
	2T	80		0.05	0.11		0.28	0.05	0.04	0.16	0.02	0.01	0.18		0.01		0.09	11	2.634

[a]Shannon–Weaver index.

TABLE 7
ADH locus: number of alleles, allelic frequencies and allelic diversity in pasture and mine populations

Soil type	Site code	Number of plants sampled (gene number observed)	Alleles (+ SE) 1.12	1.00	0.88	Number of alleles	I_s[a]
Pasture	1N	35 (140)	0.01 (0.02)	0.99 (0.02)		2	0.08
	2N	35 (140)	0.06 (0.04)	0.93 (0.04)	0.01 (0.02)	3	0.407
	3N	35 (140)	0.18 (0.06)	0.81 (0.06)	0.01 (0.009)	3	0.757
Mine	1T	60 (240)	0.28 (0.05)	0.58 (0.08)	0.14 (0.04)	3	1.367
	2T	80 (320)	0.29 (0.05)	0.60 (0.05)	0.11 (0.03)	3	1.310
	3T	110 (440)	0.42 (0.05)	0.49 (0.05)	0.09 (0.03)	3	1.343

[a]Shannon–Weaver index.

TABLE 8

GOT 1 locus: number of alleles, allelic frequencies and allelic diversity (I_s) in pasture and mine populations

Soil type	Site code	Number of plants sampled (gene number observed)	Alleles (+ SE)					Number of alleles	I_s[a]
			1.08	1.04	1.00	0.84	0.92		
Pasture	1N	52 (208)		0.07 (0.03)	0.50 (0.07)	0.43 (0.07)		3	1.293
	2N	35 (140)		0.17 (0.06)	0.43 (0.08)	0.40 (0.08)		3	1.383
	3N	35 (140)	0.10 (0.05)	0.10 (0.05)	0.69 (0.08)	0.11 (0.05)		4	1.49
Mine	1T	51 (204)			0.68 (0.06)	0.32 (0.06)		2	0.435
	2T	90 (360)			0.38 (0.05)	0.53 (0.05)	0.09 (0.03)	3	1.329
	3T	113 (452)		0.06 (0.02)	0.75 (0.04)	0.19 (0.04)		3	1.010

[a] Shannon–Weaver index.

TABLE 9

GOT 2 locus: number of alleles, allelic frequencies and allelic diversity (I_s) in pasture and mine populations

Soil type	Site code	Number of plants sampled (gene number observed)	Alleles (+ SE)							Number of alleles	I_s[a]
			1.05	1.00	0.94	0.88	0.82	0.80	0.70		
Pasture	1N	52 (208)		0.83 (0.05)	0.17 (0.05)					2	0.658
	2N	35 (140)		0.91 (0.05)	0.09 (0.05)					2	0.195
	3N	35 (140)		0.93 (0.04)	0.07 (0.04)					2	0.366
Mine	1T	51 (204)		0.93 (0.04)	0.05 (0.04)	0.01 (0.01)		0.01 (0.01)		4	0.435
	2T	90 (360)	0.02 (0.01)	0.61 (0.05)	0.28 (0.05)	0.09 (0.03)				4	1.375
	3T	113 (452)	0.02 (0.01)	0.91 (0.03)	0.04 (0.02)	0.01 (0.009)		0.01 (0.009)	0.01 (0.009)	6	0.631

[a]Shannon–Weaver index.

$$H_e = 1 - \sum_{i=1}^{n} s_i^2$$

where s_i is the allelic frequency of the ith allele and n the number of alleles studied.

The probabilities of there being heterozygote gametes from the heterozygotes *iiij, iijj, iijk, ijkl* are respectively 1/2, 2/3, 5/6 and 1, where i, j, k and l represent alleles at one locus. Therefore H_o is given by:

$$H_o = \frac{1/2\ x\ (iiij) + 2/3\ y\ (iijj) + 5/6\ z\ (iijk) + w\ (ijkl)}{N}$$

where N is the number of individuals analysed and x, y, z and w are the number of individuals in the different classes of heterozygotes.

The estimate of F includes effects due to inbreeding and population substructuring. It has not been possible to separate these effects in this study; however comparison between F values reflects the variation of the genotypic structure between populations. Positive values of F can be interpreted as the average inbreeding coefficient of a population.

B. Comparison between adjacent populations from pasture and mine

F values have been calculated for both the GOT 1 locus and the ADH locus, but not for the GOT 2 locus because of the low frequencies of all alleles except one within most of the populations. These values are given in Table 12 (see basic data, Tables 10 and 11). For locus GOT 1, high heterozygote deficits were constant in pasture populations. By contrast, mine populations were quite different from one another: a high deficit was detected at site 2T, almost none was found for site 1T, and a slight excess occurred at site 3T. The same variation appeared for locus ADH in mine populations, but F values cannot be taken into account in pasture populations because of the high frequency of the 1.00 allele (more than 81%). Thus, no correlation appears between F values and the two types of populations.

When the F values are plotted against the number of flowering plants per square metre, it is clear that the F parameter increases with increasing density (Fig. 4). A linear regression analysis of variance confirms that the increase in heterozygote deficit with increase in panicle density is highly significant for GOT 1 ($P < 0.001$), and a clear trend is obtained with ADH locus.

C. Within-population variation

In order to confirm the density effect we eliminated the habitat effect. For this purpose a complementary study was carried out in a calaminary meadow (site 1)

TABLE 10
Genotypic frequencies at locus GOT 1 in pasture and mine populations

Soil type	Site code	Genotypes *iiii*	*iiij*	*iijj*	*iijk*	*ijkl*	*N*
Pasture	1N	0.50	0.29	0.13	0.08		52
	2N	0.89	0.11				35
	3N	0.52	0.37	0.11			35
Mine	1T	0.25	0.45	0.30			51
	2T	0.59	0.24	0.12	0.05		90
	3T	0.25	0.49	0.10	0.16		113

TABLE 11
Genotypic frequencies at locus ADH in pasture and mine populations

Soil type	Site code	Genotypes *iiii*	*iiij*	*iijj*	*iijk*	*ijkl*	*N*
Pasture	1N	0.97		0.03			35
	2N	0.86		0.08	0.06		35
	3N	0.68	0.03	0.29			35
Mine	1T	0.52	0.16	0.32			60
	2T	0.58		0.40	0.02		80
	3T	0.21	0.12	0.58			110

TABLE 12
Locus GOT 1 and locus ADH: *F* values[a] within the three pairs of adjacent populations

Soil type	Site code	GOT 1 locus number of plants sampled	*F*	ADH locus number of plants sampled	*F*	Panicles m^{-2}
Pasture	1N	52	0.47		[b]	61
	2N	35	0.90		[b]	352
	3N	35	0.48		[b]	85
Mine	1T	51	0.03	60	0.44	15
	2T	90	0.49	80	0.49	69
	3T	113	−0.11	110	0.22	1.6

[a]See basic data for *F* value calculation: Tables 10 and 11.
[b]*F* values cannot be taken into account because of the high frequency of the 1.00 allele (> 0.81).

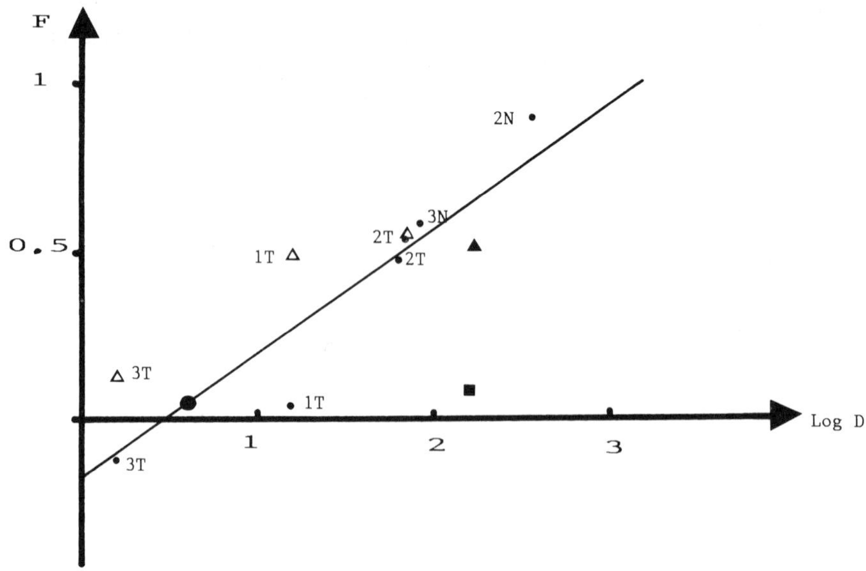

FIG. 4 Correlation between *F* value and density (*D*). Small dots (.) represent data for the locus GOT 1 from populations of the three sites (1,2,3). The regression line is given for seven samples because site 1T has been sampled twice ($y = 0.36 \log D - 0.16$, $r = 0.97$, $p = 0.001$). Open triangles (Δ) represent data for the locus ADH (populations of toxic soils only). Filled markers represent data for locus GOT 1 from a population of the calaminary meadow, high density area (central part: ▲, peripherical part ■), low density area (●).

by comparing the genetic structure in the high density area (average density: 100 panicles per m^2; 152 individuals sampled in the central part and 62 in the peripheral part) and a surrounding low density area (average density: 2 panicles per m^2; 80 individuals sampled).

The expected density effect was found at the GOT 1 locus: the central and dense zone had a strong deviation from panmixia, whereas the non-dense zone and the peripheral part of the dense zone both approached the Hardy–Weinberg equilibrium (see Fig. 4). It should be emphasized that populations growing on mine soil are usually less dense than those occurring on non-toxic soils.

As far as the infrequent alleles 0.84, 1.04 and 1.08 are concerned, a structuring of the allele distribution was observed. In almost all characters studied a higher level of genetic diversity was found in the mine population (Table 13). The *F* value is usually high in pastures, which are most frequently dense, and contributes to reducing the genotypic diversity in these populations.

V. Conclusions

The results from this study fully confirm the data of Wu *et al.* (1975). They also lead to the same conclusion: types tolerant to extreme habitats are not necessarily

TABLE 13

Level of genetic diversity in adjacent populations of *Arrhenatherum elatius* (− = low; + = high; 0 = no differences observed).

Characteristics	Pasture	Mine
Physiology and morphological traits		
Flowering time	−	+
Plant height	−	+
Node phenotype	−	+
Isozyme variation		
TO system	−	+
Allozyme variation		
Allelic diversity		
locus ADH	−	+
locus GOT 1	0	0
locus GOT 2	−	+
Genotypic diversity (locus ADH and GOT 1)	− (very often = heterozygote deficit due to density effect)	+ (very often: panmixia in sparse populations)

rare in normal pastures, and no genetic drift resulting from a more or less unique founder effect is perceptible.

The *higher* genetic diversity on mine soils may be due either to selective effects or to neutral effects. Among the selective forces, diversity could be the consequence of the spatial heterogeneity which is found in many habitats perturbed by man. Mine habitats correspond to a large range of ecological niches. This explanation is well supported by the pronounced small-scale variation in heavy metal content of the mine soils, contrasting with the homogeneity of pasture soils (Fig. 2). Moreover, increased diversity of mine habitats could result from a low level of biotic selection. If the selective forces generated by edaphic factors seem to be stronger in the mine area than in the pasture area (fewer species observed in the mine area and usually a low density of *Arrhenatherum elatius* populations, see Table 1), at the other extreme the low intensity of biotic selective forces might favour the maintenance of more pronounced diversity. Conversely, in pasture habitats selection originates mainly from biotic competition and acts strongly upon various characters such as height, growth and potential for using habitat resources. A unique founder event could generate the reverse situation, i.e. an increased diversity. This possibility cannot be rejected because of the more rapid population turnover observed on mine soils. Higher diversity might also result from a more open breeding system. Although the mating system and its variations were not directly studied, variation in the genotypic structure gives interesting clues. The high heterozygote deficit in dense populations may be considered as an inbreeding effect, gene flow being more limited in dense populations than in sparse colonies, since individuals in dense populations are more likely to mate with their relatives.

The relationship between the density and the genetic structure of population may be explained in two ways:

- an increase of the actual selfing rate in the high density area. However, in *Arrhenatherum elatius* the level of autofertility is relatively low, not exceeding 5% (Pfitzenmeyer, 1962). In our experiment (Ducousso, 1985) the selfing rates in mine pasture populations were comparable (2.08% versus 2.27%).
- a reduction of the size of the genetic neighbourhood by a limited dispersal of the seed and pollen in a high density area. It is conceivable that an average individual from a dense population would be pollinated preferentially by its immediate neighbours. Its offspring grow near the mother plant, and after several generations a high rate of effective inbreeding appears. On the contrary, an individual from a scattered population is more likely to be pollinated on average by pollen from more distant and more numerous unrelated donors, and the population remains at the Hardy–Weinberg equilibrium.

Ellstrand *et al.* (1978) have shown that dense populations of *Helianthus annuus* have lower outcrossing rates than less dense populations. This result was presumably due to the presence of intrafamily inbreeding in the dense population. Such a pattern of change in the size of the genetic neighbourhood has been described in other species, e.g. *Primula vulgaris* (Cahalan and Gliddon, 1985). This phenomenon, possibly not uncommon in many natural populations, is very important because it may lead to genetic differentiation. Wright (1969) considers that genetic differentiation due to random forces is possible if the size of the neighbourhood does not exceed 200; on the other hand, it is very likely that if the neighbourhood size remains smaller than 20 small subpopulations with a non-random distribution will be formed. In this case, a Wahlund's effect (1928) due to sampling appears, reinforcing the deficit of heterozygotes.

It seems that the higher genetic diversity of mine populations, which are very often not dense, could also partly result from the higher levels of gene flow between individuals from these populations.

The main points of our results can be summarized as follows:

(1) no clear-cut differences in the genetic structure between pairs of populations were found, but if the heterozygote deficit is plotted against the population density a highly suggestive relationship is obtained: the highest heterozygote deficit is detected in the more dense population;

(2) the heterozygote deficit is interpreted as a consanguinity effect originating from a more limited gene flow in dense populations than in sparse populations;

(3) morphological characters as well as allozyme markers indicate that the genetic diversity of populations inhabiting toxic soils is higher than that of populations

occurring in normal pastures. This suggests that tolerant types must not be very rare in normal pasture populations from which tolerant populations have evolved;

(4) The higher genetic diversity of populations from toxic stations might be related to: (i) the spatial heterogeneity of toxic habitats; (ii) the higher sensibility toward gene flow of individuals from these populations which trap pollen from greater distances than do plants from dense populations; (iii) the low intensity of biotic selective forces.

References

Antonovics, J. (1968). Evolution in closely adjacent plant population. V. Evolution of self fertility. *Heredity* **23**, 219–238.

Bazzaz, F. A. (1983). Characteristics of populations in relation to disturbance in natural and modified ecosystems. *In* "Disturbance and Ecosystems" (H. A. Mooney and M. Godron, eds), pp. 259–275. Springer, Berlin.

Brown, A. D. H. (1979). Enzyme polymorphism in plant populations. *Theor. Pop. Genet.* **61**, 315–320.

Cahalan, C. M. and Gliddon, C. (1985). Genetic neighbourhood in *Primula vulgaris. Heredity* **54**, 65–70.

Caisse, M. and Antonovics, J. (1978). Evolution in closely adjacent plant populations. IX. Evolution of reproductive isolation in clinal populations. *Heredity* **40**, 371–384.

Davis, B. J. (1964). Disc electrophoresis. II. Method and application to human serum proteins. *Ann. N.Y. Acad. Sci.* **121**, 404–427.

Ducousso, A. (1985). Comparaison de populations adjacentes d'*Arrhenatherum elatius* sur des sols normaux et toxiques: analyse de la diversité génétique et due régime de la reproduction. Thesis, University of Lille.

Ducousso, A., Petit, D., Valero, M. and Vernet, Ph. (Submitted). Genetic variation between and within populations of a perennial grass, *Arrhenatherum elatius. Heredity.*

Ellstrand, N. C. and Marshall, D. L. (1985). The impact of domestication on distribution of allozyme variation within and among cultivars of radish, *Raphanus satiuus* L. *Theor. Appl. Genet.* **69**(4), 393–398.

Ellstrand, N. C., Torres, A. M. and Levin, D. A. (1978). Density and the rate of apparent outcrossing in *Helianthus annuus* (Asteraceae). *Syst. Bot.* **3**, 403–407.

Ennos, R. A. (1984). The mating system and genetic structure in a perennial grass, *Cynosurus cristatus* L. *Heredity* **55**, 121–126.

Gasquez, J. and Compoint, J. P. (1976). Apport de l'électrophorèse en courant pulsé à la taxonomie d'*Echinochloa crus-galli* L. *Ann. Amélior.* **26**, 345–355.

Gouyon, P. H. and Vernet, Ph. (1982). The consequences of gynodioecy in natural populations. *Theor. Appl. Genet.* **61**, 315–320.

Gouyon, P. H., Lumaret, R., Valdeyron, G. and Vernet, Ph. (1983). Reproductive strategies and disturbance by man. *In* "Disturbance and Ecosystems" (H. A. Mooney and M. Godron, eds), pp. 214–225. Springer, Berlin.

Hamrick, J. L., Linhart, Y. B. and Mitton, J. B. (1979). Genetic variation and life history traits. *Ann. Rev. Ecol. Syst.* **10**, 173–200.

Jain, S. K. (1976). The evolution of breeding in plants. *Ann. Rev. Ecol. Syst.* **7**, 469–495.

Kimura, M. and Maruyama, T. (1971). Pattern of neutral polymorphism in a geographically structured population. *Gent. Res. Camb.* **18**, 125–131.

Ledig, F. T. and Conkle, M. T. (1983). Gene diversity and genetic structure in a narrow endemic, Torrey Pine (*Pinus Torreyana* Parry ex. Carr.). *Evolution* **37**, 371–384.

Levin, D. A. (1976). Consequences of long-term artificial selection, inbreeding and isolation in *Phlox*. 2. The organisation of allozymic variability. *Evolution* **30**, 463–472.

Levin, D. A. (1978). Genetic variation in annual *Phlox*: self-compatible species. *Evolution* **32**, 245–263.

Lumaret, R. (1981). Structure génétique d'un complexe polyploide *Dactylis glomerata* L.. Thesis, U.S.T.L. Montpellier.

Lumaret, R. and Valdeyron, G. (1980). Les dactyles de la Crau: mise en évidence de relations entre différents écotypes par le polymorphisme enzymatique. *C.R. Acad. Agric.* **66**, 229–238.

Maynard-Smith, J. (1978). "The Evolution of Sex". Cambridge University Press, Cambridge.

McNeilly, T. and Antonovics, J. (1968). Evolution in closely adjacent plant populations. IV. Barriers to gene flow. *Heredity* **40**, 371–384.

Nevo, E. (1978). Genetic variation in natural populations: patterns and theory. *Theor. Pop. Biol.* **13**, 121–177.

Orstein, B. J. (1964). Disc electrophoresis. I. Background and theory. *Ann. N.Y. Acad. Sci.* **121**, 321–347.

Pfitzenmeyer, C. D. (1962). Biological flora of the British Isles: *Arrhenatherum elatius* (L.) J. et C. Presl. *J. Ecol.* **50**, 235–245.

Shannon, C. E. and Weaver, W. (1963). "The Mathematical Theory of Communication". Urbana University Press, Illinois.

Solbrig, O. T. (1972). Breeding system and genetic variation in *Leavenworthia. Evolution* **26**, 155–160.

Valdeyron, G., Dommée, B. and Valdeyron, A. (1973). Gynodioecy: another computer simulation model. *Am. Nat.* **107**, 454–459.

Valero, M. and Olivieri, I. (1985). Adjacent populations of cocksfoot (*Dactylis glomerata* L.): a detailed study of allozyme varation across contrasting habitats. *In* "Genetic Differentiation and Dispersal in Plants" (P. Jacquard, G. Heim and J. Antonovics, eds), pp. 339–354. NATO ASI Series, Springer, Berlin.

Verleij, J. A. C., Bast-Cramer, W. B. and Levering, H. (1985). Effect of heavy metal stress on the genetic structure of populations of *Silene cucubalus. In* "Structure and Functioning of Plant Populations", Vol. 2 (J. Haeck and J. W. Woldendorp, eds), pp. 355–365. North-Holland, Amsterdam.

Wahlund, S. (1928). Zusammensetzung von Populationen und Korrelationserscheinungen vom Standpunkt der Vererbungslehre aus betrachtet. *Hereditas* **11**, 65–106.

Wright, S. (1969). Evolution and genetics of population. *In* "Theory of Gene Frequencies", Vol. 2. University of Chicago Press, Chicago.

Wu, L., Bradshaw, A. D. and Thurman, D. A. (1975). The potential for evolution of heavy metal tolerance in plants. III. The rapid evolution of copper tolerance in *Agrostis stolonifera. Heredity* **34**, 165–185.

8 Demographic Differentiation and Life-history Evolution in Temperate Woodland Plants

SCHOICHI KAWANO[1], TAKENORI TAKADA[2], SHUICHI NAKAYAMA[1] and AKIRA HIRATSUKA[3]

[1]*Department of Botany, Faculty of Science, Kyoto University, Japan*
[2]*Department of Biophysics, Faculty of Science, Kyoto University, Japan*
[3]*Biological Institute, Faculty of Science, Tohoku University, Sendai, Japan*

I. Introduction

Life-history studies of the populations of a single plant species or species group have been to a large extent concerned with evolutionary–ecological radiation within a group, the consequences of adaptive responses to the environments, or selection and fitness for particular ecological conditions. The differentiation of diverse gender types, complex pollination and breeding systems and perennation strategies, including various types of asexual reproduction, are unique features of plants. They are undoubtedly related to the sessile nature of plants and the autotrophic system that captures and fixes solar radiation.

Since 1968, we have been carrying out comparative studies on the life-history characteristics of temperate woodland herbs (Kawano *et al.*, 1968, 1982; Kawano, 1970, 1975, 1985; Kawano and Nagai, 1975). This chapter reports some of our results, the emphasis being on the demographic features of four herb species representative of temperate woodlands in Japan.

Classical demographic methods focus on the dynamics of age-structured populations, using techniques originally developed for studies of animal populations. In plant species, however, individuals of the same age are usually in very

Differentiation Patterns in Higher Plants
ISBN: 0-12-709425-3

different demographic states, and age-classified demography cannot be directly applied for dealing with such complex life histories. A number of authors who have studied the demography of plant populations have already adapted the population projection matrix technique of Lefkovitch (1965; see also Leslie 1945, 1948) to the complex life cycles of herbaceous plants and trees (Hartshorn, 1975; Werner and Caswell, 1977). In a series of recent papers, Caswell (1978, 1982) presented general formulae for the stable *stage* distribution — the analogue of the stable *age* distribution of classical demography — and the stage-specific reproductive value distribution for populations with complex life histories.

In the present study we have attempted to apply a similar technique to four temperate woodland herbs with complex life histories. This method proved to be useful for analysing the demographic structure of plant populations in the field. We also present a model describing the ecological–evolutionary trends in demographic patterns of various temperate woodland herbs.

II. Life-history Characteristics of Four Liliaceous Species

A. Erythronium japonicum

Various life-history parameters of four liliaceous species selected for the present study are compiled and compared with each other in Tables 1 and 2. *Erythronium japonicum* is a typical polycarpic perennial, its life span extending over 20 years. It attains sexual maturity in approximately six to seven years. Flowers are hermaphroditic, insect-pollinated (especially by *Xylocopa* and other large social bees), and obligately xenogamous (Kawano *et al.*, 1982). In short, *E. japonicum* is a typical sexually reproducing species, and regeneration takes place almost exclusively by seeds. Exceptionally new sprouts arise from pieces of the bulb-appendage broken by mechanical pressures (Kawano, 1982).

In order to study the demography of *E. japonicum*, ten permanent plots (1 m quadrats) were established in the natural habitats of this species. Every single individual was carefully marked, mapped, and measurements of leaf size were taken over the eight-year period since 1979. Sixteen different size-classes or stage-classes were distinguished in this study. Fourteen sterile stage-classes, including the seedling stage, and ten fertile stage-classes, were distinguished in relation to the leaf area or biomass of individual plants (Table 3). This classification constituted a convenient key for further analysis of natural populations. Figure 1 shows an example of mapping of individual plants (quadrat no. 9). Their size-classes were determined first, and the positions were recorded for each individual.

Studies on population flux included all important life-history parameters, namely net change in plant numbers, rate of increase, survival and mortality rates, expected time for complete turnover, etc. (Table 4). The number of individuals representing different stage-classes varied greatly in the plots studied, as exemplified by Fig. 2. In the plot Q 1 the population is not saturated, and a gradual

TABLE 1

Life-history parameters in four liliaceous species of temperate woodlands

Taxa	Life span (years)	Age of first reproduction (years)	Reproductive life span	Sexuality	Breeding system	Pollination system
Erythronium japonicum	Perennial, > 20	*c.* 7	Polycarpic	Hermaphrodite	Xenogamous	Insects
Disporum sessile	Pseudo-annual,[a] < 1	< 1	Monocarpic	Hermaphrodite	Xenogamous (or partially autogamous?)	Insects
Disporum smilacinum	Pseudo-annual, < 1	< 1	Monocarpic	Hermaphrodite	Xenogamous (or partially autogamous?)	Insects
Allium monanthum	Pseudo-annual, < 1	< 1	Monocarpic	Dioecious (or rarely hermaphrodite and andro-monoecious)	Xenogamous	Insects

[a]Salisbury (1942).

TABLE 2
Reproductive characteristics in the four species studied

Taxon	No. of flowers/ plant	No. of female gametes/ plant	No. of male gametes/ plant	Male/ female (P/O)	No. of seeds/ plant	Fecundity	No. of vegetative propagules/ plant
Erythronium japonicum[a]	1	38.7 ± 10.6 (18–77)	1.6–2.4 × 10^5	4,100–6,150	25.0 ± 12.3	66.4 ± 29.7	—
Disporum sessile[b]	1–3	6–18	1.5–4.5 × 10^4	*c*.25,000	4.7 ± 2.6	< 39.0	1.0–1.8
Disporum smilacinum[b]	1–2(–5)	6–12(–30)	5.9–11.2(–29.4) × 10^4	9,300–9,800	1.5 ± 0.7	< 25.0	2.3–2.8
Allium monanthum (Female)[c]	1–2	6–12	—	1,091–12,450	2–4	< 25.0	1.0–4.0
(Male)	—	—	1.3–7.5 × 10^4	—	—	—	1.0–3.0

[a]Kawano *et al*. (unpubl.).
[b]Kawano and Hiratsuka (unpubl.).
[c]Kawano and Nagai (1975).

TABLE 3

Erythronium japonicum: size-classes based on the leaf areas of individuals[a]

Size-class	Sterile		Fertile	
	Range (cm^2)	Mean ± SD	Range (cm^2)	Mean ± SD
Class 0 (Seedling)				
1	0.58– 1.43	0.97 ± 0.29	—	—
2	1.52– 2.90	2.26 ± 0.49	—	—
3	3.03– 5.68	4.42 ± 0.85	—	—
4	6.12– 9.87	7.98 ± 1.13	—	—
5	10.13– 19.82	13.95 ± 3.17	—	—
6	20.58– 29.42	23.51 ± 2.76	—	—
7	31.90– 39.22	35.08 ± 2.61	—	—
8	41.29– 49.10	45.55 ± 3.11	47.91– 49.13	48.52 ± 4.65
9	51.24– 61.76	58.53 ± 4.65	50.44– 63.57	59.26 ± 4.24
10	66.18– 78.46	71.71 ± 4.43	65.17– 79.73	70.23 ± 4.42
11	90.21– 99.53	95.21 ± 3.99	81.12– 98.32	88.24 ± 5.56
12	101.11–113.61	107.18 ± 4.11	101.30–115.70	106.55 ± 5.39
13	121.01–133.23	126.67 ± 5.43	125.23–132.03	130.04 ± 3.25
14	—	—	144.44–165.64	151.59 ± 7.64
15	—	—	177.60–198.08	186.88 ± 8.41
16	—	—	205.71–230.98	219.01 ± 12.69

[a]Kawano *et al*. (1982).

TABLE 4
Population flux in *Erythronium japonicum* sites (each site 1 m^2)

	Q 1	Q 2	Q 4	Q 5	Q 6	Q 7	Q 8	Q 9	Q 10
(a) No. of plants/m^2, April 1979	33	8	26	33	65	80	93	142	37
(b) No. of plants/m^2, April 1986	82	8	22	46	65	64	85	149	47
(c) Net change (b−a)	49	0	−4	13	0	−16	−9	7	10
(d) Rate of increase (b/a)	2.49	1.00	0.85	1.39	1.00	0.80	0.89	1.05	1.27
(e) No. of plants arrived between April 1979 and April 1986	96	5	52	39	62	50	97	177	62
(f) Total no. of plants lost between April 1979 and April 1986	38	5	32	21	62	41	97	150	42
(g) Plants present April 1979, alive by April 1986	20	5	7	23	54	52	49	76	17
(h) Percentage survival of plants in (a) (g/a × 100)	60.6	62.5	26.9	69.7	83.1	65.0	51.0	53.5	46.0
(i) Expected time for complete turnover (year) (100×7/(100-h))	17.8	18.6	7.00	19.8	41.4	20.0	14.4	15.1	13.0
(j) Total plants recorded during study	122	12	54	67	127	125	182	299	89
(k) Percentage annual mortality of all individuals (f/j × 100)	0.31	0.42	0.59	0.31	0.49	0.33	0.53	0.50	0.47

increase in the number of individuals is still taking place. On the other hand, the population of the plot Q 9 is nearly saturated, there being more than 300 plants per m^2. We noticed considerable annual fluctuations in the numbers of seedlings, reflecting the changes in seed production in different years.

Changes in net population size, cumulative gains and losses of individuals were also different in the two plots selected (Fig. 3). Curiously enough, the replacement of individuals by newly born ones occurs more frequently in the saturated population (Q 9) than in the non-saturated one (Q 1). Percentages of flowering individuals per plant above the critical size also fluctuate drastically every season, no doubt affecting the levels of fecundity for each local population.

B. Disporum sessile *and* D. smilacinum

Both *Disporum sessile* and *D. smilacinum* are pseudo-annual in the sense of Salisbury (1942) and monocarpic. Their dry matter economy is of a typical annual type: all reserved food in the underground stock organs is entirely consumed for producing new aerial shoots, and new photosynthate produced is stocked in new ramets normally formed in different underground positions. All the ramets produced become completely separate and independent at the end of every season (Kawano and Nagai, 1975; Kawano, 1984, 1985); they are xenogamous and insect-pollinated.

In *Disporum* species there are no runners at the beginning of the season, but during the season one to three (rarely five) underground runners are produced, and at the tip of the runners small buds for the next season are formed (Fig. 4). However, since the mother plant finally decays at the end of the season, all ramets produced at the tip of each runner become separate and independent (Fig. 5). While seed production also takes place, the rate of sexual offspring recruitment is severely limited.

The size-class structures of *Disporum sessile* populations were surveyed at Yatsuo, Toyama Prefecture (Fig. 6). In this species, the size-classes distinguished were based on the number of leaves borne per plant. There is a conspicuous correlation between the biomass and number of leaves, reflecting the growth stage (Kawano and Hiratsuka, unpubl.). We have sampled many populations, but the patterns observed were very consistent, intermediate size-classes always being the best represented.

The population structures in *Disporum smilacinum* (Fig. 7) are very similar to those in *D. sessile*, intermediate classes being numerous. Mature plants are confined to the larger size-classes. Seasonal changes in the proportion of plants with different numbers of underground runners observed from April to December were comparable in *D. sessile* and *D. smilacinum* (Fig. 8). In April, all plants lacked runners, but the proportion of plants with runners gradually increased, and at the end of the season in December all ramets became separate and independent. This process of growth and reproduction was confirmed

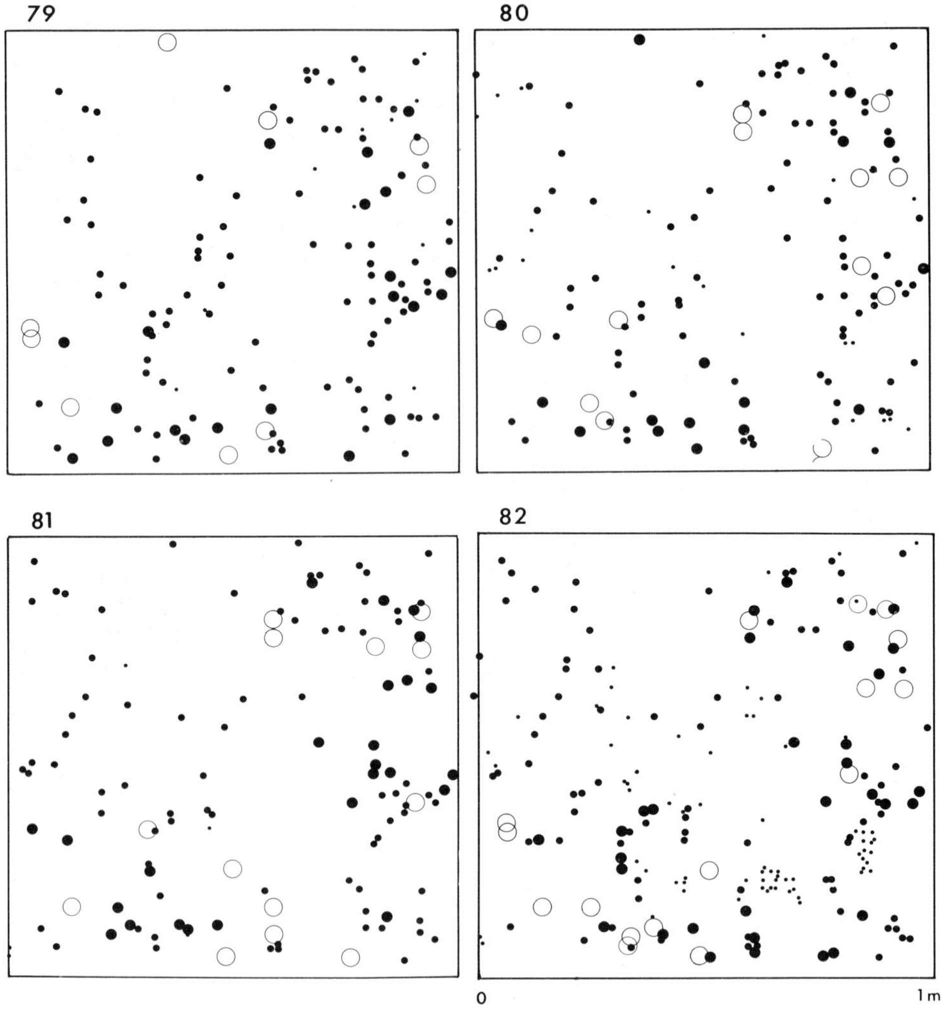

FIG. 1(a) Distribution and seasonal changes of *Erythronium japonicum* individuals over the four years (1979–82) in the plot Q 9 at Yatsuo, Toyama Prefecture. Large open circles, flowering individuals; large filled circles, sterile individuals belonging to size-classes 8–13; medium filled circles, sterile size-classes 1–7; small filled circles, seedlings.

experimentally (Fig. 9). Small size-classes normally produce large size-classes, although the rate is low. Intermediate size-classes produce the same intermediate size-classes, and rarely produce larger size-classes. In contrast the large size-classes always produce smaller, intermediate size-classes.

The complex offspring recruitment in *Disporum sessile* and *D. smilacinum* results in odd population structures always possessing the highest number of

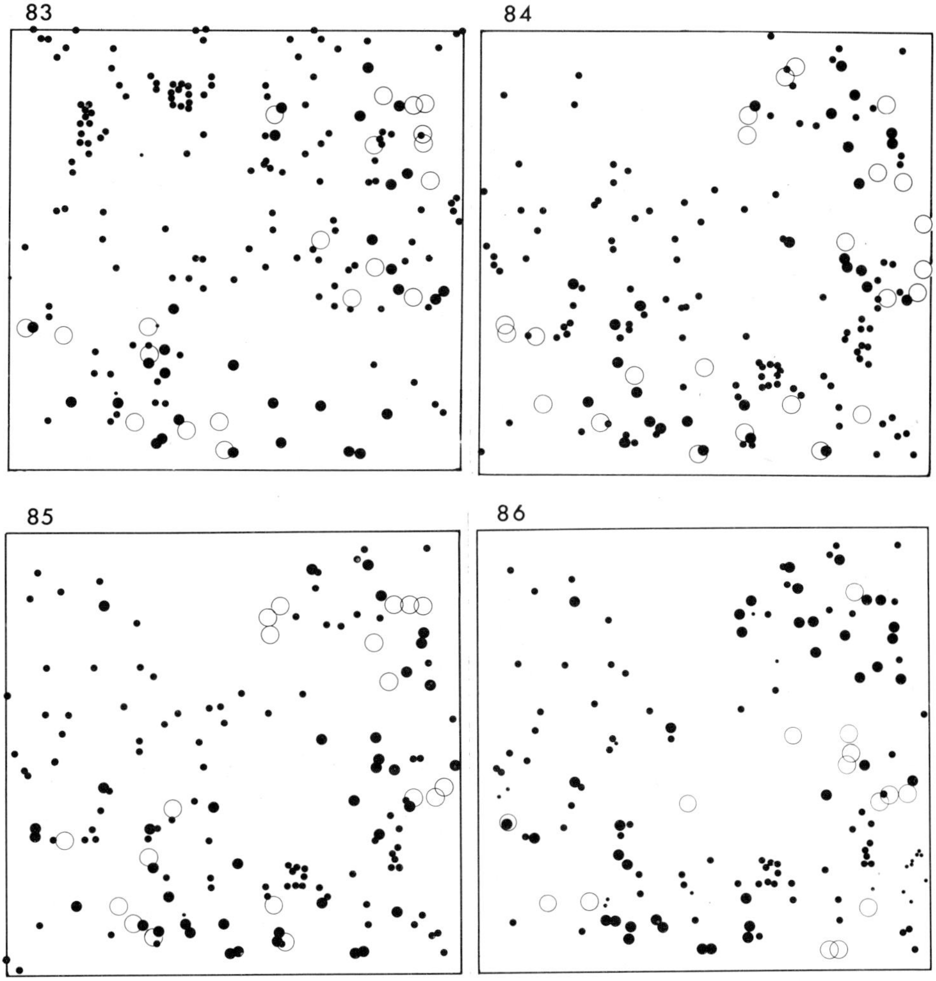

FIG. 1(b) Distribution and seasonal changes of *Erythronium japonicum* individuals over a further four years (1983–86) in the plot Q 9 at Yatsuo. See Fig. 1(a) for details.

intermediate size-classes among the populations. This fact is perhaps more easily understood if we recall the reproductive capacity of these two species. The seed output per plant is exceedingly low, attaining on average only one to several seeds per plant. The offspring recruitment is therefore almost exclusively dependent on asexual reproduction. This fact is also well documented by the data on the genetic structures of local populations, obtained by using peroxydase isozyme polymorphisms as genetic markers (Kawano, 1984; Kawano *et al.*, unpubl.).

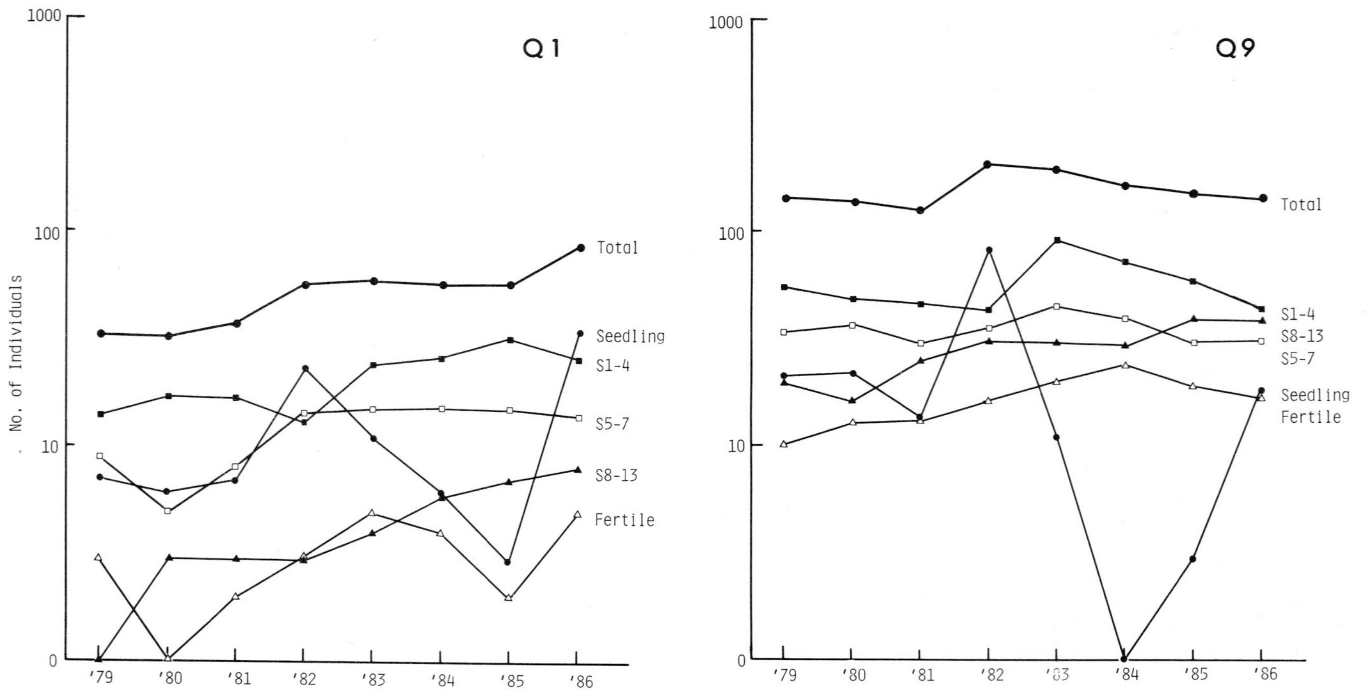

FIG. 2 Seasonal changes in number of *Erythronium japonicum* individuals belonging to different stage-classes in the plots Q 1 and Q 9.

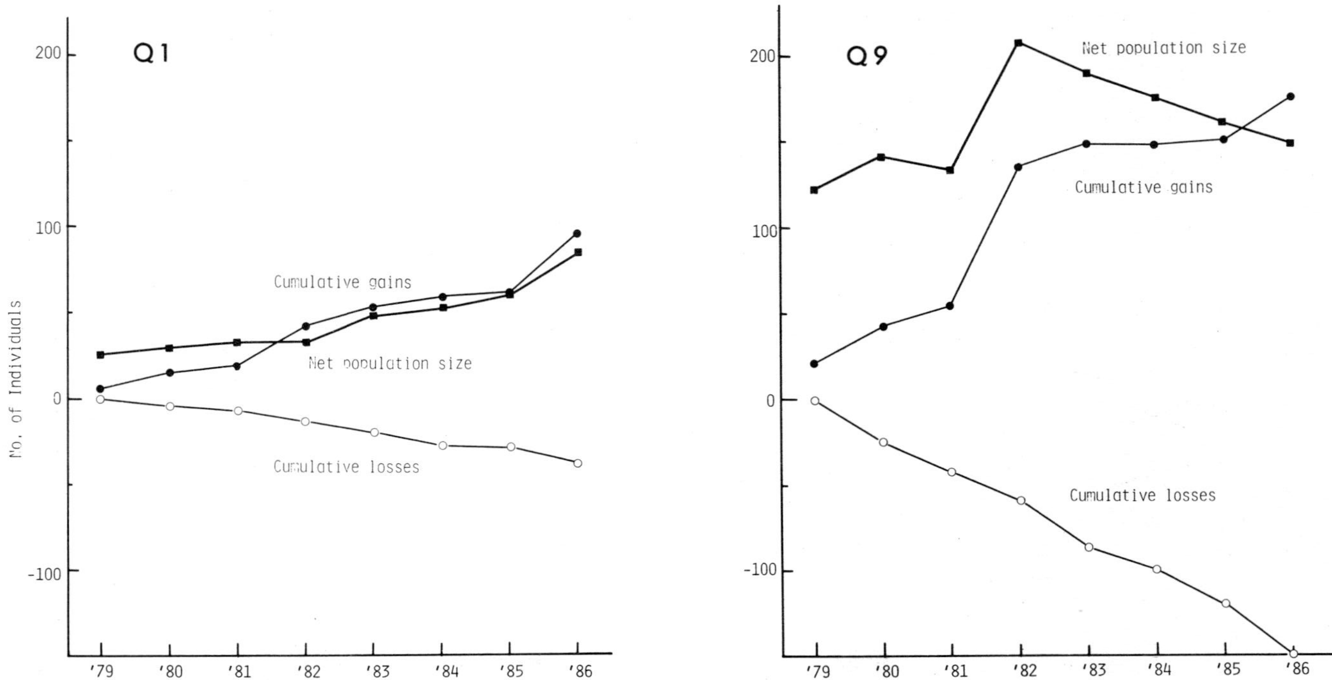

FIG. 3 Seasonal changes in net population size, cumulative gains and losses of *Erythronium japonicum* in the plots Q 1 and Q 9 (see Table 4).

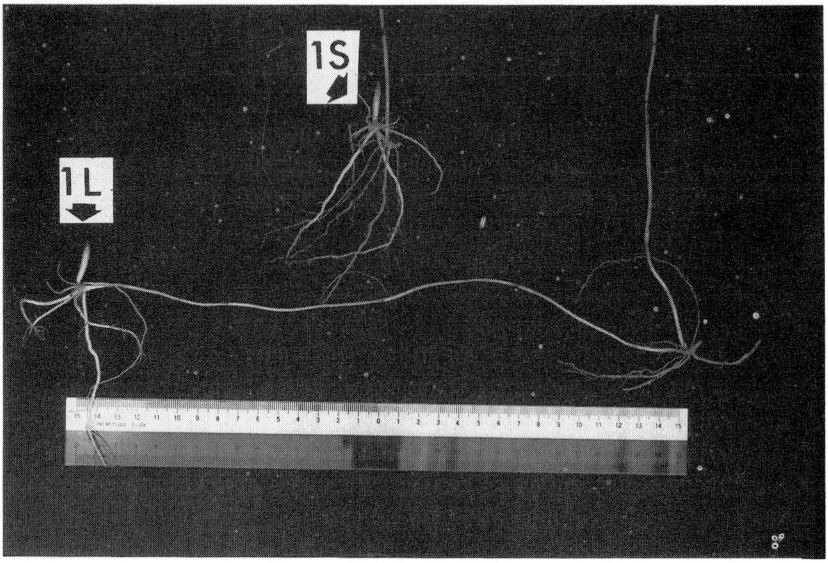

FIG. 4 Root system of *Disporum smilacinum*. 1S, apical bud produced at the tip of a short underground runner; 1L, apical bud produced at the tip of a long underground runner.

C. Allium monanthum

This pseudo-annual species is characterized by a most unique reproductive strategy. It is a tiny, typical spring plant about 10 cm tall, which remains above ground for only two weeks in early spring. As its specific epithet indicates, it is dioecious (or rarely andromonoecious, Fig. 10); curiously enough, male plants are extremely rare, and have been found only in a few localities in Japan (Kawano and Nagai, 1975, unpubl.). The chromosome numbers and karyotypes in *A. monanthum* are diversified. We found diploids ($2n = 16$), triploids ($2n = 24$), tetraploids ($2n = 32$), and in addition several structural hybrids (Noda and Kawano, unpubl.).

The reproductive system of *A. monanthum* is exceedingly complex and comprises six different reproductive types described below (Nagai, 1972; Kawano and Nagai, 1975). 1L and 2L are plants with a single leaf or two leaves, respectively (Fig. 10). The food reserve in the old bulb is entirely consumed every season, but new photosynthate is stocked in the new bulbs formed in the same position. 1LS and 2LS are those with a single leaf or two leaves, but with a slender creeping underground stolon. The old bulb (marked with an arrow) disappears, owing to the complete consumption of food reserve, and new bulbs are formed just beneath the leaf. As a result, new offspring individuals in these reproductive types migrate at least 20 cm or more every season. 1LF and 1LF are basically the same as 1L and 2L, but with a single scape. Plants may be male, female, hermaphrodite, or andromonoecious, but those with male sexuality are extremely rare (Kawano and Nagai, 1975, unpubl.).

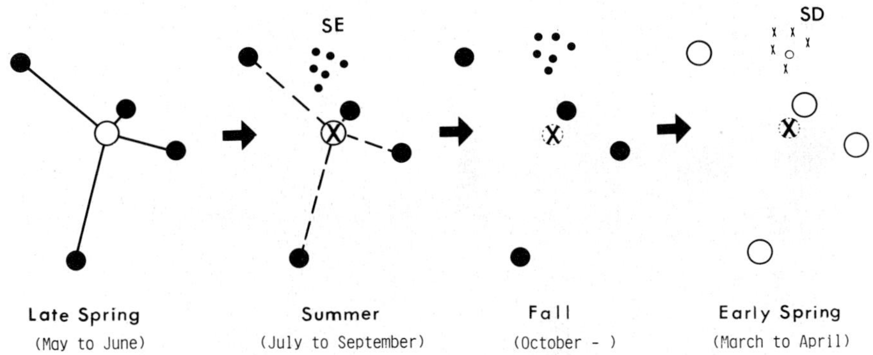

FIG. 5 Ramet and seed production and dispersal in *Disporum* species. Large open circles, mother plant; large filled circles, ramets (or vegetative propagules) produced at the tip of underground runners; small filled circles (SE), sexual propagules; small open circles (SD), juvenile plants derived from sexual propagule. X indicates the death of individuals.

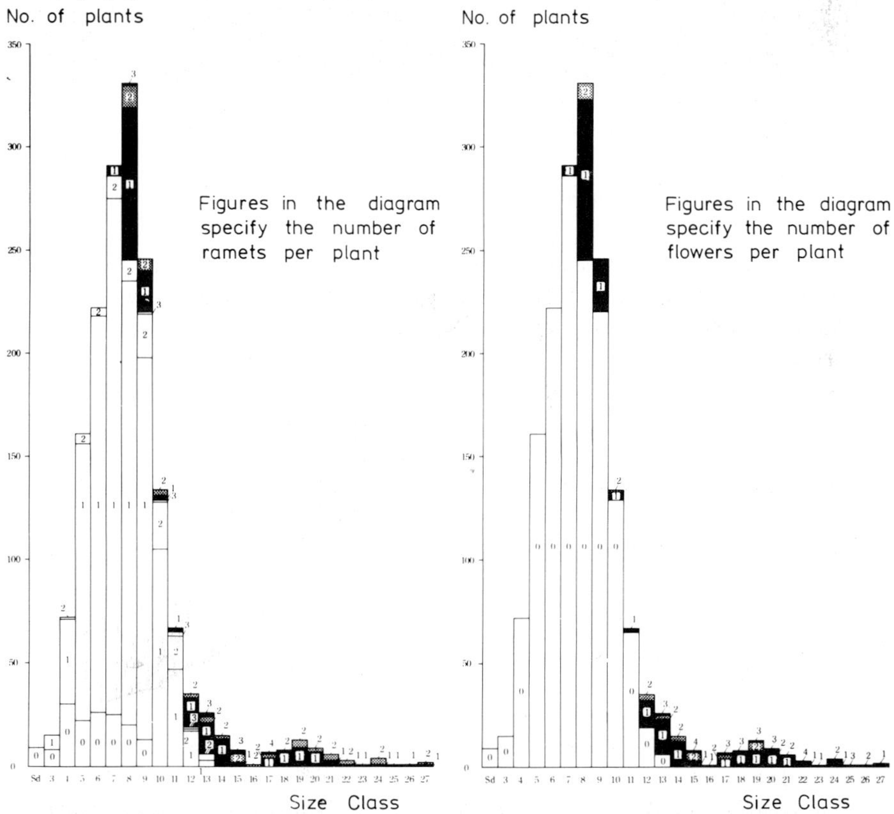

FIG. 6 Size-class structure of *Disporum sessile* sampled from one 2.5 m × 2.5 m quadrat on the deciduous forest floor in Yatsuo, Toyama Prefecture. Sd, seedlings, 3–27, different size-classes based on the number of leaves per plant.

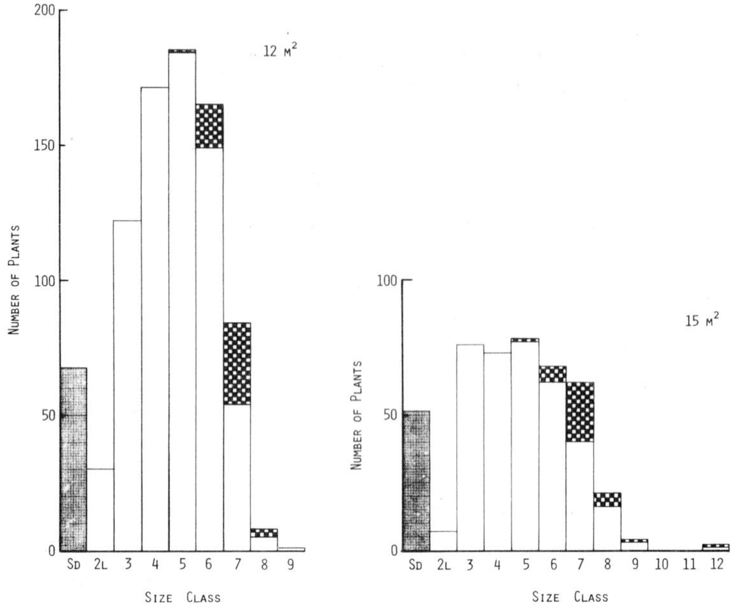

FIG. 7 Size-class structure of *Disporum smilacinum* sampled from two sites (twelve 1 m × 1 m quadrats and fifteen 1 m × 1 m quadrats, randomly scattered) on the deciduous forest floor in Yatsuo, Toyama Prefecture. The number of seeds per plant (Sd) in the two populations (12 × 1 m^2 and 15 × 1 m^2) is estimated as the mean number of seed produced per plant multiplied by the number of flowering plants. 2L−9L, eight size-classes based on the number of leaves per plant.

Vegetative propagation in *A monanthum* followed a very specific pattern, three different reproductive types originating from three daughter bulblets formed within a single large bulb (Fig. 11). Two plants newly formed from daughter bulblets in the outer part had stolons, whereas one plant formed in the middle had no stolons. As a result, the three offspring plants originating from a large mother bulb migrated, occupying different positions. The cultivation experiments demonstrate that in any reproductive type, both 1L and 1LS types are most frequently produced in the next season (Table 5).

III. Comparison of Demographic Structures between Four Woodland Species

A. Lefkovitch matrix

In order to analyse further the demographic patterns and characteristics of four woodland species, we provide for each species a Lefkovitch matrix (Lefkovitch, 1965), based on the stage-class distribution data. Lefkovitch proposed a population model describing the change with time of the number of individuals for size stages in the organism's life cycle.

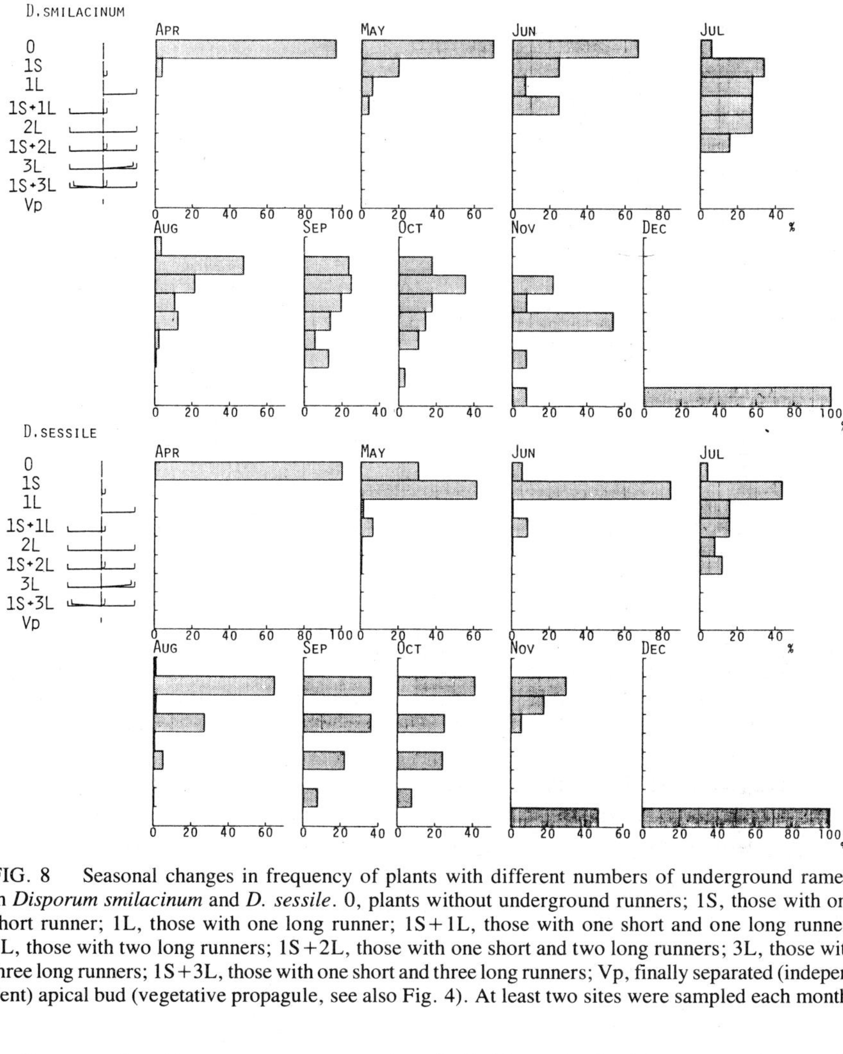

FIG. 8 Seasonal changes in frequency of plants with different numbers of underground ramets in *Disporum smilacinum* and *D. sessile*. 0, plants without underground runners; 1S, those with one short runner; 1L, those with one long runner; 1S+1L, those with one short and one long runner; 2L, those with two long runners; 1S+2L, those with one short and two long runners; 3L, those with three long runners; 1S+3L, those with one short and three long runners; Vp, finally separated (independent) apical bud (vegetative propagule, see also Fig. 4). At least two sites were sampled each month.

Suppose that there are p stages and $n_{i,t}$ individuals in the ith stage at time $t+1$; then $n_{i,t+1}$ can be written:

$$n_{i,t+1} = a_{i1}n_{1,t} = a_{i2}n_{2,t} + \ . \ . \ . + a_{ip}n_{p,t}$$
$$= \Sigma_j a_{ij}n_{j,t} \qquad (1)$$

where a_{ij} is an element of the Lefkovitch matrix and represents the contribution of an individual in the jth stage to the number of individuals in the ith stage at time $t+1$. The relation between $n_{i,t+1}$ and $n_{i,t}$ for all stages is expressed as

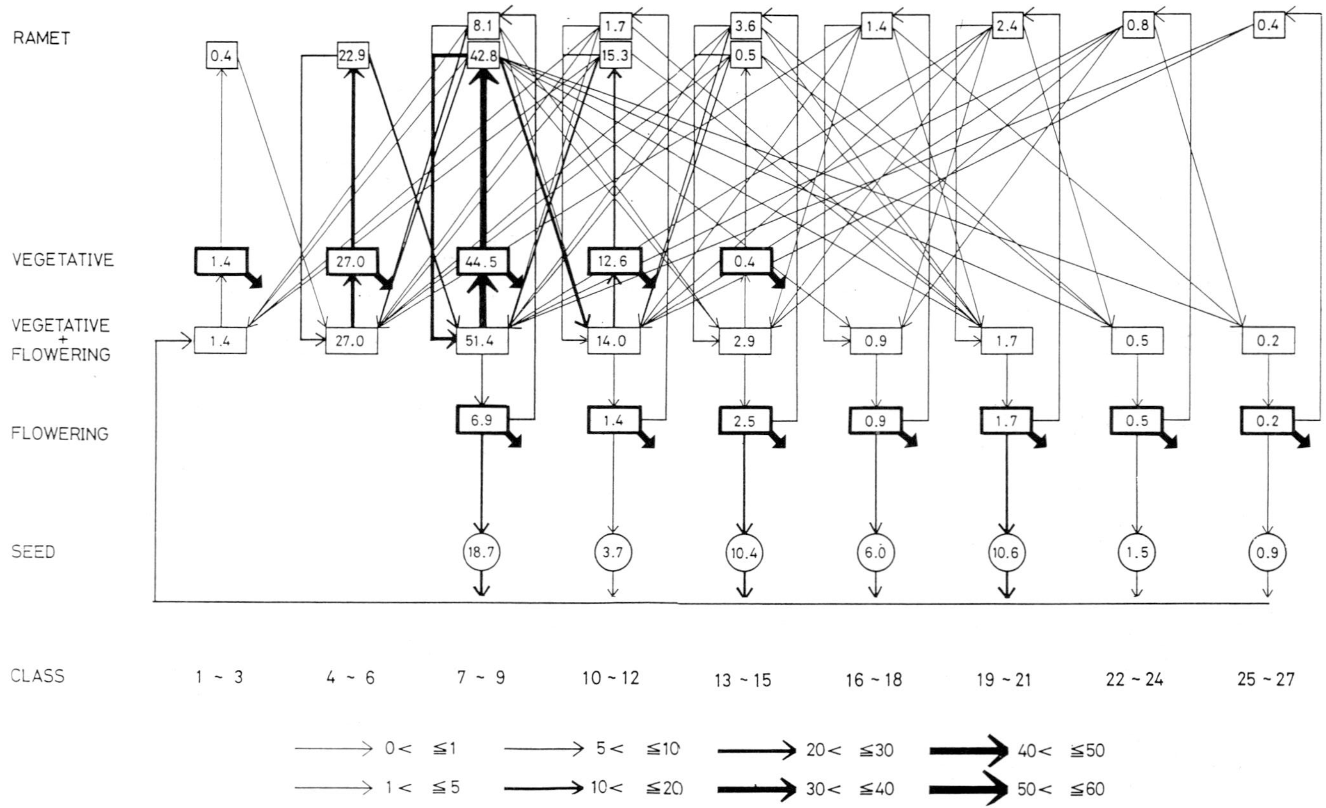

FIG. 9 The rates of origin of each size-class from its parent individual in *Disporum sessile*. The horizontal axis specifies size-classes based on the number of leaves per plant. Arrows indicate the directions of sexual or asexual propagule formation. The figures and thickness of arrows show the birth rate of certain size-classes as percentages. All the parental plants of vegetative and flowering stages decay at the end of the season (shown in thick boxes). The figures given in the circles and boxes indicate the number of seeds and/or vegetative propagules (expressed here as percentages of the total).

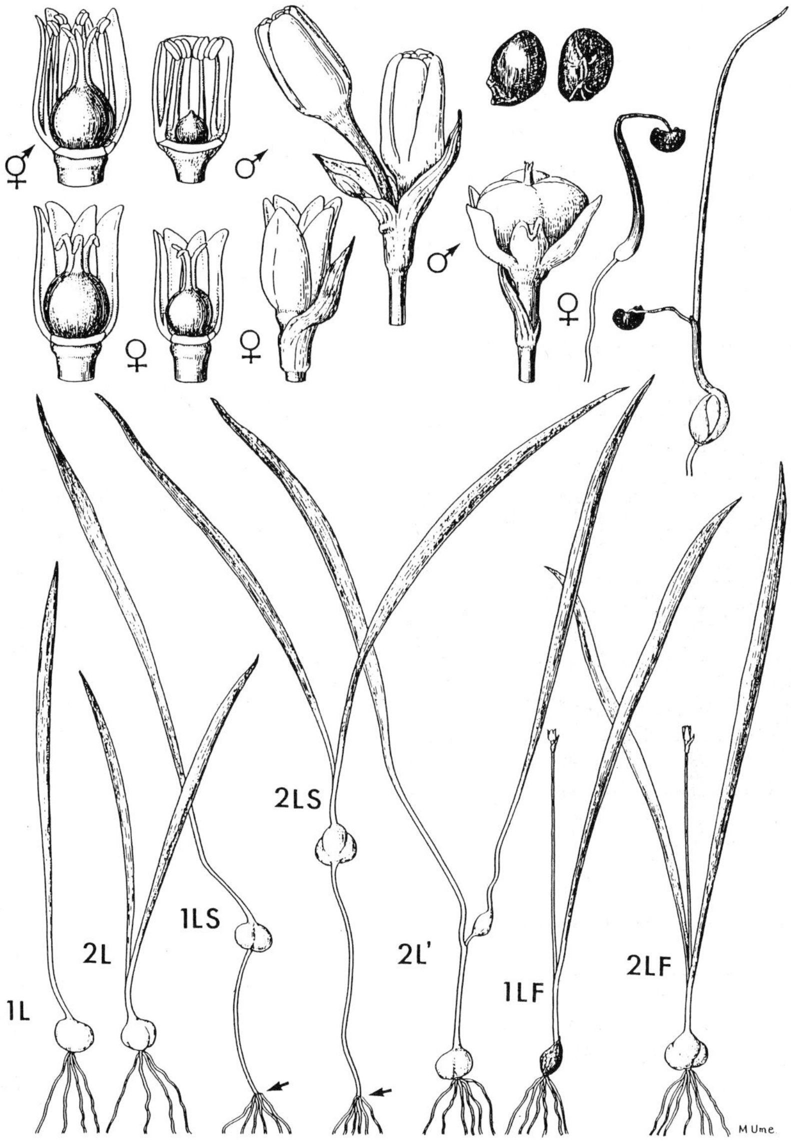

FIG. 10 Sexuality and varying vegetative reproductive types in *Allium monanthum*.

$$\begin{pmatrix} n_1 \\ n_2 \\ \vdots \\ n_p \end{pmatrix}_{t+1} = \begin{pmatrix} a_{11} & a_{12} & \cdots & a_{1p} \\ a_{21} & a_{22} & \cdots & a_{2p} \\ \vdots & & & \vdots \\ a_{p1} & a_{p2} & \cdots & a_{pp} \end{pmatrix} \begin{pmatrix} n_1 \\ n_2 \\ \vdots \\ n_p \end{pmatrix}_t \tag{2}$$

or

$$\mathbf{n}_{t+1} = A\mathbf{n}_t \tag{3}$$

where $\mathbf{n}_t$ is a vector representing the stage composition of the population $N_t = \Sigma_i n_{i,t}$.

Figure 12 shows the Lefkovitch matrix for *Erythronium japonicum*. The horizontal axis indicates the size-classes of the current year (SE, SD, . . ., F14–16) and the vertical axis indicates those of the succeeding year. First we look at the sterile stage, $a_{SD,SE}$ (0.09) being the transitional probability from seed to seedling, i.e., the survival rate of seed. Each component of the second row of the matrix is the rate of transition from seedling to the subsequent sterile stages: 0.17 to the S1 stage, 0.27 to the S2 stage, 0.13 to the S3 stage, and so on. The contribution to a succeeding year of each stage, $\Sigma_i a_{ij}$, is shown by the dotted region in Fig. 12. Since *E. japonicum* has no vegetative reproduction, the values of the sterile stages are less than 1.

We now look at the flowering stage. The seed supply from the *i*th stage, i.e. the number of seeds produced per plant, is coded by $a_{SE,i}$. These values are generally greater than 1. The contribution to the next generation by means of sexual reproduction is shown by the hatched region in Fig. 12. In the case of *E. japonicum*, sexual reproduction plays the most significant role for the recruitment of individuals to the next generation, although accidental vegetative reproduction resulting from breakage of bulb-appendages is also known (Kawano, 1982).

The Lefkovitch matrix of *Disporum sessile* is shown in Fig. 13. The meaning of each element of the matrix is the same as in the case of *E. japonicum*. However, it should be noted that in a pseudo-annual species like *Disporum*, vegetative reproduction plays the most important role in the recruitment of offspring. The values greater than 1 in the bottom histogram indicate the efficient productivity by both sexual and asexual reproduction. Although *D. sessile* and *D. smilacinum* (Fig. 14) are both pseudo-annual species, the dependency of offspring recruitment by asexual reproduction in the former species is much lower than in the latter. In *D. sessile*, the proportion of sexual reproduction in the contribution to the succeeding year is $2.71/\Sigma_i a_{i,F3} = 70\%$ at size-class F3; however, in *D. smilacinum* it is $0.96/\Sigma_i a_{i,F3} = 35\%$ at size-class F3.

In *Allium monanthum* (Fig. 15) such a tendency is remarkably strong. In this unique pseudo-annual species, offspring recruitment depends almost exclusively on asexual reproduction.

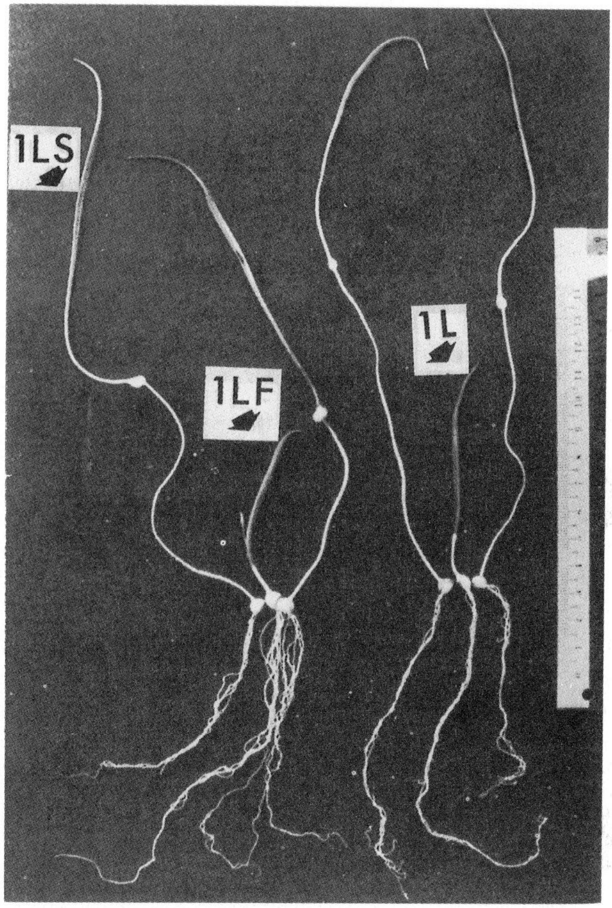

FIG. 11 Reproductive types derived from the 2LF (female) type in the succeeding season. Left, two 1LS and one 1LF types produced from three daughter bulblets; right, two 1LS and one 1L types produced from three daughter bulblets (after Kawano and Nagai, 1975).

B. Flow diagram

Here we attempt to compare the Lefkovitch matrices of all four species at once, four significant life-history stages, namely seed, juvenile, adult (sterile) and adult (fertile), being taken into consideration. Adults (S) includes all individuals exceeding a critical size and having the potential for reproduction. For example, in the case of *Erythronium japonicum*, juvenile stages include SD and S1 – 8. Adult (S) consists of S9 – 14 and adult (F) of F9 – 16. These aggregated Lefkovitch matrices are shown in Table 6.

Based on these matrices, we can obtain the steady-state frequency vector for each species, which satisfies the following equation:

TABLE 5

The origin of various reproductive types in *Allium monanthum*

Original type	Types produced								Total
	1L (2L′)	1LS (2LS′)	2L	2LS	1LF (F)	2LF (M)	1LF (F)	2LF (M)	
1L	314	226	49	4	1	3	—	—	597
(%)	(52.6)	(37.9)	(8.2)	(0.7)	(0.2)	(0.5)	—	—	
1LS	229	205	5	—	—	—	(1)	(1)	441
(%)	(51.9)	(46.5)	(1.1)	—	—	—	(0.2)	(0.2)	
2L	91	176	3	3	—	—	—	(1)	274
(%)	(51.9)	(46.5)	(1.1)	(1.1)	—	—	—	—	
2LS	22	43	—	—	—	—	—	—	65
(%)	(33.9)	(66.1)	—	—	—	—	—	—	
1LF (F)	35	27	2	2	1	1	—	—	68
(%)	(51.5)	(39.7)	(2.9)	(2.9)	(1.5)	(1.5)	—	—	
2LF (F)	53	101	4	—	—	—	—	—	158
(%)	(33.4)	(63.9)	(2.5)	—	—	—	—	—	
1LF (M)	8	2	—	—	—	—	—	6	16
(%)	(50.0)	(12.5)	—	—	—	—	—	(37.5)	

F, female; M, male.

	SE	SD	S1	S2	S3	S4	S5	S6	S7	S8-10	S11-13	F8-10	F11-13	F14-16
SE	---	---	---	---	---	---	---	---	---	---	---	*18.87*	*26.92*	*25.20*
SD	0.09	---	---	---	---	---	---	---	---	---	---	---	---	---
S1	---	0.17*	0.29*	0.11*	0.03	0.01	---	---	---	---	---	---	---	---
S2	---	0.27*	0.17*	0.27*	0.07	0.02	0.01	---	---	---	---	---	---	---
S3	---	0.13*	0.06	0.38*	0.46*	0.09	0.01	0.01	---	---	---	---	---	---
S4	---	0.03	0.03	0.03	0.25*	0.31*	0.06	0.01	0.01	---	0.01	---	---	---
S5	---	---	0.01	0.02	0.05	0.46*	0.51*	0.05	0.03	0.01	0.01	0.02	---	---
S6	---	---	0.01	0.01	0.01	0.01	0.29*	0.38*	0.12*	0.02	0.01	0.02	---	---
S7	---	---	---	---	---	0.01	0.03	0.32*	0.14*	0.02	---	0.04	---	---
S 8-10	---	---	---	---	---	---	0.02	0.15*	0.59*	0.54*	0.11*	0.63*	0.20*	---
S11-13	---	---	---	---	---	---	---	0.01	---	0.12*	0.20*	0.11*	0.31*	0.13*
F 8-10	---	---	---	---	---	---	0.06	0.01	0.03	0.16*	0.10*	0.11*	0.08	---
F11-13	---	---	---	---	---	---	---	0.01	0.02	0.10*	0.52*	0.07	0.37*	0.47*
F14-16	---	---	---	---	---	---	---	---	---	---	0.02	---	0.01	0.40*

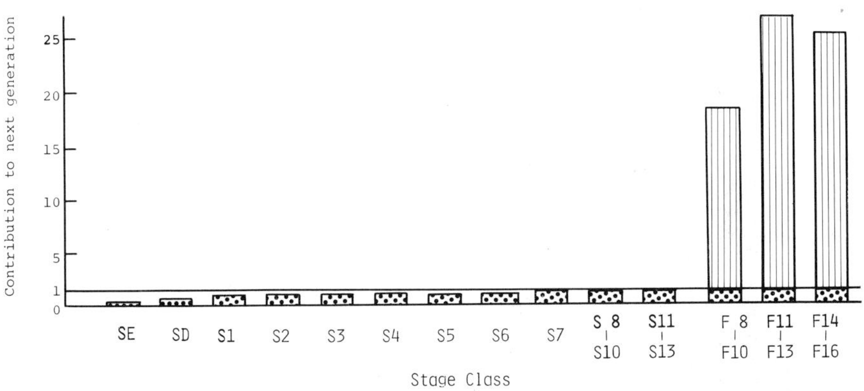

FIG. 12 The Lefkovitch matrix of *Erythronium japonicum*. a_{ij} = the rate of transition from jth stage-class to ith stage-class: italics denote the number of seeds produced per plant, i.e. the offspring recruitment by sexual reproduction. Elements with "*" have values exceeding 0.1.

	SE	V1	V2	V3	V4	V5	F3	F4	F5	F6	F7	F8	F9
SE	---	---	---	---	---	---	*2.71*	*2.64*	*4.15*	*6.67*	*6.24*	*3.00*	*4.50*
V1	0.10*	---	---	0.01	0.03	---	0.05	0.24*	---	---	---	---	---
V2	---	0.29*	0.28*	0.18*	0.03	0.21*	0.76*	0.24*	0.14*	---	---	---	---
V3	---	---	0.49*	0.31*	0.83*	0.18*	0.23*	0.21*	0.13*	0.23*	0.20*	0.28*	0.87*
V4	---	---	---	0.32*	0.14*	0.19*	0.05	0.22*	0.65*	0.23*	0.21*	0.29*	0.90*
V5	---	---	---	---	---	0.02	0.01	---	0.02	0.04	0.03	0.04	---
F3	---	---	0.08	0.05	0.13*	0.03	0.04	0.03	0.02	0.04	0.03	0.04	0.13*
F4	---	---	---	0.04	0.02	0.02	0.01	0.02	0.07	0.03	0.02	0.03	0.10*
F5	---	---	---	0.01	---	0.18*	0.04	---	0.13*	0.23*	0.20*	0.28*	---
F6	---	---	---	0.01	---	---	---	---	---	0.26*	0.24*	0.32*	---
F7	---	---	---	0.01	0.03	0.21*	---	0.24*	0.14*	0.26*	0.24*	---	---
F8	---	---	---	0.01	---	0.21*	---	---	0.14*	---	0.24*	---	---
F9	---	---	---	0.01	---	---	---	---	---	0.26*	---	0.32*	---

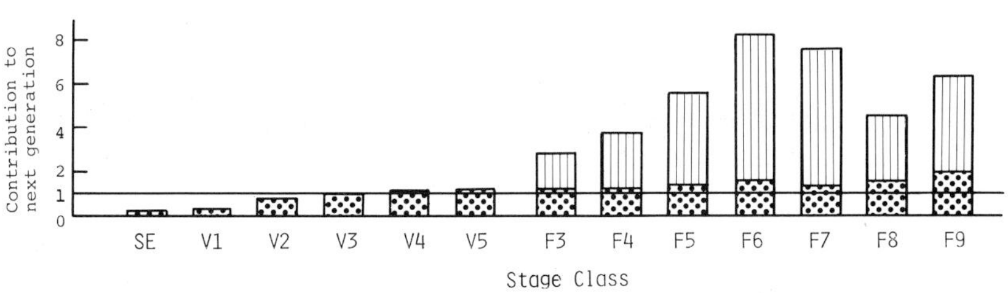

FIG. 13 The Lefkovitch matrix of *Disporum sessile*. The meaning of each element in the matrix is the same as in Fig. 12. Size-classes from F3 to F9 represent the flowering stages, those from V1 to V5 the vegetative growth stages.

	SE	V1	V2	V3	F3	F4	F5	F6
SE	---	---	---	---	0.96	1.60	1.00	1.00
V1	0.11*	0.47*	---	---	0.17*	0.21*	---	---
V2	---	0.47*	0.53*	0.53*	0.18*	0.21*	0.75*	---
V3	---	---	0.43*	0.72*	1.03*	1.03*	0.62*	1.65*
F3	---	---	0.09	0.15*	0.22*	0.22*	0.13*	0.35*
F4	---	---	---	0.18*	0.18*	---	0.75*	---
F5	---	---	---	---	0.01	---	---	---
F6	---	---	---	0.01	0.02	---	---	---

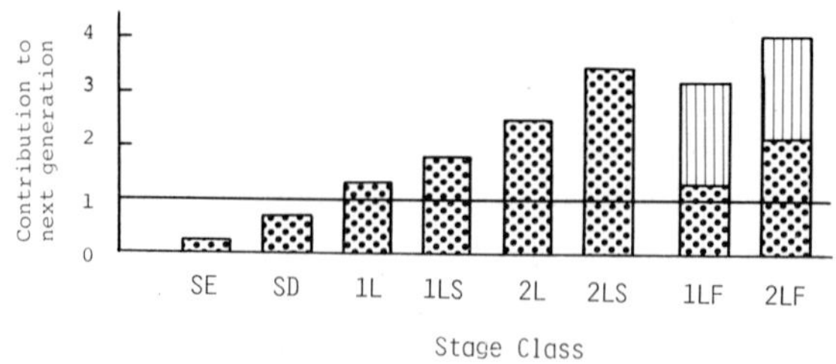

FIG. 14 The Lefkovitch matrix of *Disporum smilacinum*. The meaning of each element in the matrix is the same as in Fig. 12. Size-classes from F3 to F6 represent the flowering stages, those from V1 to V3 the vegetative growth stages.

	SE	SD	1L	1LS	2L	2LS	1LF	2LF
SE	---	---	---	---	---	---	1.88	1.88
SD	0.18*	---	---	---	---	---	---	---
1L	---	0.70*	0.70*	0.92*	0.84*	1.17*	0.69*	0.73*
1LS	---	---	0.51*	0.82*	1.62*	2.29*	0.53*	1.40*
2L	---	---	0.11*	0.02	0.03	---	0.04	0.06
2LS	---	---	0.01	---	0.02	0.01	---	---
1LF	---	---	---	---	---	---	0.02	---
2LF	---	---	0.01	---	0.03	---	0.04	---

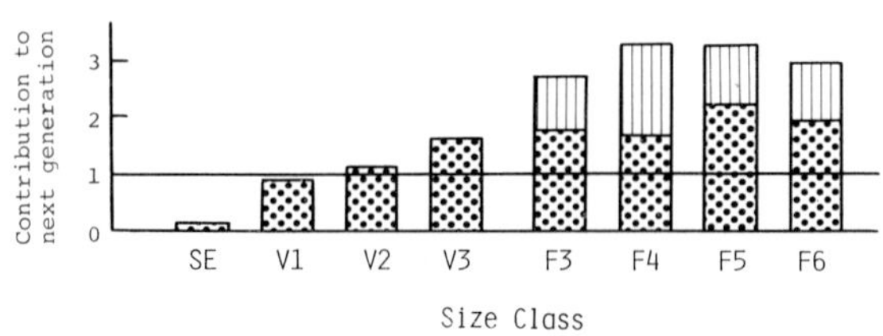

FIG. 15 The Lefkovitch matrix of *Allium monanthum*. The meaning of each element in the matrix is the same as in Fig. 12. Size-classes from 1LF to 2LF represent the flowering stages, those from 1L to 2LS the vegetative growth stages.

$$\lambda \mathbf{f} = B\mathbf{f} \tag{4}$$

where λ is an eigenvalue of B representing the Malthusian parameter of the population, $\mathbf{f}$ is a steady-state frequency vector and B is an aggregated Lefkovitch matrix:

$$\mathbf{n}_{t+1} = B\mathbf{n}_t \tag{5}$$

or

$$n_{i,t+1} = \Sigma_i b_{ij} n_{j,t}$$

TABLE 6
The Lefkovitch matrices of the four species studied

(a) *Erythronium japonicum*

$i+1/i$	Seed	Juvenile	Adult(S)	Adult(F)
Seed	—	—	—	24.982
Juvenile	0.093	0.752	0.049	0.270
Adult(S)	—	0.039	0.561	0.571
Adult(F)	—	0.005	0.374	0.387

(b) *Disporum sessile*

$i+1/i$	Seed	Juvenile	Adult(S)	Adult(F)
Seed	—	—	—	3.674
Juvenile	0.100	0.281	0.163	0.430
Adult(S)	—	0.539	0.704	0.453
Adult(F)	—	—	0.172	0.319

(c) *Disporum smilacinum*

$i+1/i$	Seed	Juvenile	Adult(S)	Adult(F)
Seed	—	—	—	1.025
Juvenile	0.113	0.562	0.528	0.368
Adult(S)	—	0.481	0.974	1.061
Adult(F)	—	—	0.390	0.370

(d) *Allium monanthum*

$i+1/i$	Seed	Juvenile	Adult(S)	Adult(F)
Seed	—	—	—	1.875
Juvenile	0.180	—	—	—
Adult(S)	—	0.700	1.745	2.075
Adult(F)	—	—	0.012	0.006

Each element of $\mathbf{f}$, f_i, is the frequency of each stage class ($\Sigma_i f_i = 1$) when sufficient time has elapsed:

$$f_i = n_{i,t}/N_t \tag{6}$$

From (4), (5) and (6), $\mathbf{n}_{t+1} = \lambda \mathbf{n}_t$. Thus we obtain the following equation when sufficient time has elapsed:

TABLE 7
The flow matrices of the four species studied

(a) *Erythronium japonicum*

$i+1/i$	Seed	Juvenile	Adult(S)	Adult(F)
Seed	—	—	—	0.736
Juvenile	0.064	0.175	0.002	0.008
Adult(S)	—	0.009	0.029	0.017
Adult(F)	—	0.001	0.019	0.011

(b) *Disporum sessile*

$i+1/i$	Seed	Juvenile	Adult(S)	Adult(F)
Seed	—	—	—	0.344
Juvenile	0.033	0.050	0.065	0.040
Adult(S)	—	0.096	0.282	0.042
Adult(F)	—	—	0.069	0.030

(c) *Disporum smilacinum*

$i+1/i$	Seed	Juvenile	Adult(S)	Adult(F)
Seed	—	—	—	0.150
Juvenile	0.010	0.166	0.246	0.054
Adult(S)	—	0.142	0.454	0.155
Adult(F)	—	—	0.182	0.054

(d) *Allium monanthum*

$i+1/i$	Seed	Juvenile	Adult(S)	Adult(F)
Seed	—	—	—	0.013
Juvenile	0.001	—	—	—
Adult(S)	—	0.001	1.719	0.014
Adult(F)	—	—	0.012	0.001

$$N_{t+1} = \lambda N_t \tag{7}$$

The individual flow from stage-class i to class j when sufficient time has elapsed, can be expressed by $a_{ij}f_j$. The flow matrix having the (i,j)th element, $c_{ij} = a_{ij}f_j$, is shown in Table 7. The sum of all the elements of the flow matrix is equal to:

$$\lambda = \Sigma_i \, \Sigma_j \, c_{ij}$$

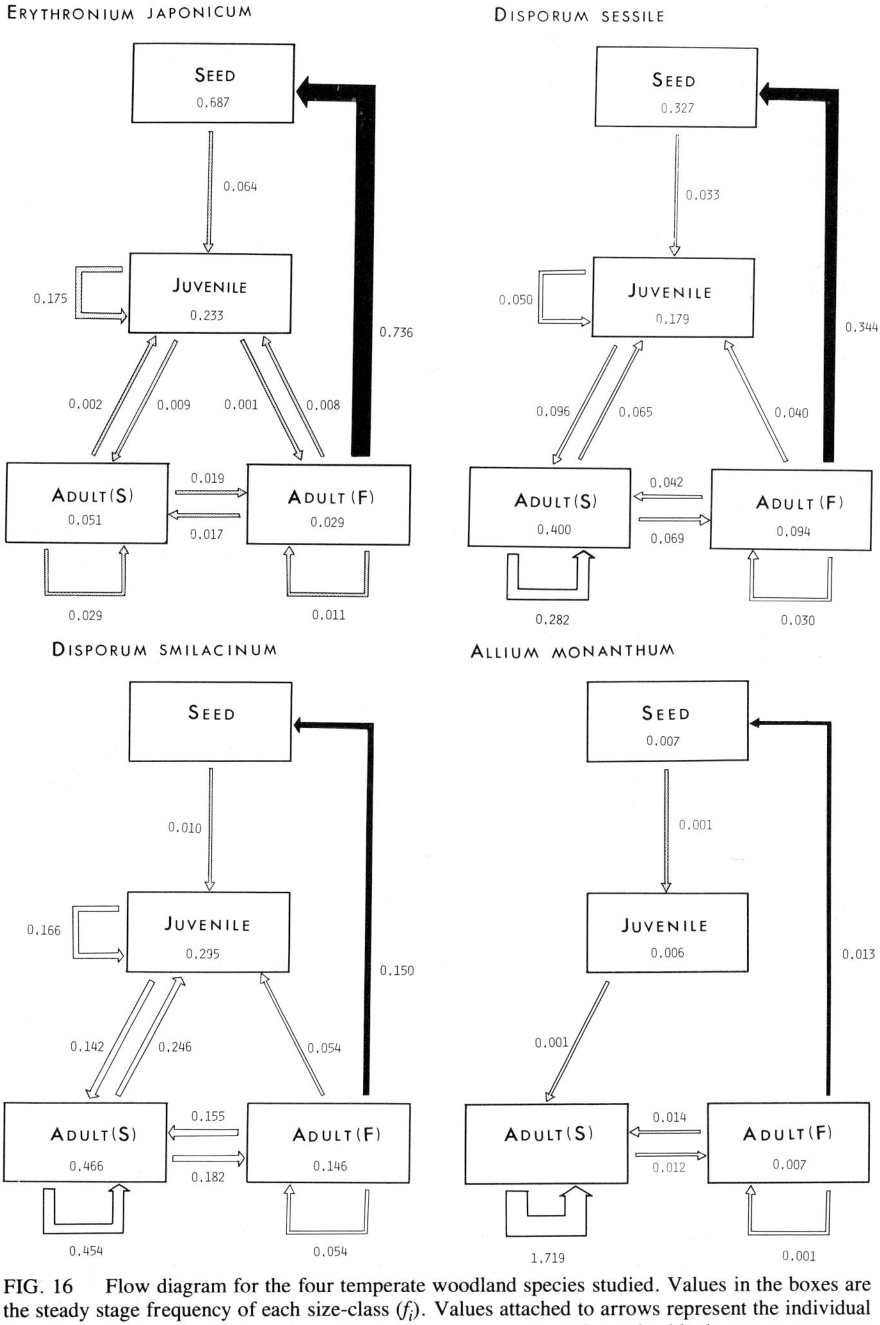

FIG. 16 Flow diagram for the four temperate woodland species studied. Values in the boxes are the steady stage frequency of each size-class (f_i). Values attached to arrows represent the individual flow from jth stage to ith stage (c_{ij}). Black arrows represent seed supply; blank arrows represent the offspring recruitment and growth through vegetative reproduction.

Dividing both sides of (5) by N_t and substituting (7) into it, we obtain the following equation:

$$\lambda f_i = \Sigma_j c_{ij}, \quad \text{i.e. } \lambda = \Sigma_i \Sigma_j c_{ij}$$

because $\Sigma f_i = 1$. Figure 16 shows the flow diagram for each species based on the flow matrix. The offspring recruitment in *E. japonicum*, a typical polycarpic perennial, is very distinct from the remaining three pseudo-annual species (Fig. 16). One significant point here is that the extent of recruitment by sexual reproduction gradually decreases from *D. sessile* to *D. smilacinum* and to *Allium monanthum*, whereas the extent of asexual propagation increases. In *A. monanthum*, the flow from adult (F) to seed is 0.011, which is one fourteenth as small as in the case of *D. smilacinum*. This implies that the population can be maintained in the absence of sexual reproduction.

C. Population growth

A computer simulation based on the Lefkovitch matrix data was attempted. Starting from 1000 seeds, the temporal increase of $\mathbf{n}_t$ is obtained by iterative calculation of (3). The time change of the population size, $N_t = \Sigma_i n_{i,t}$, is plotted on a

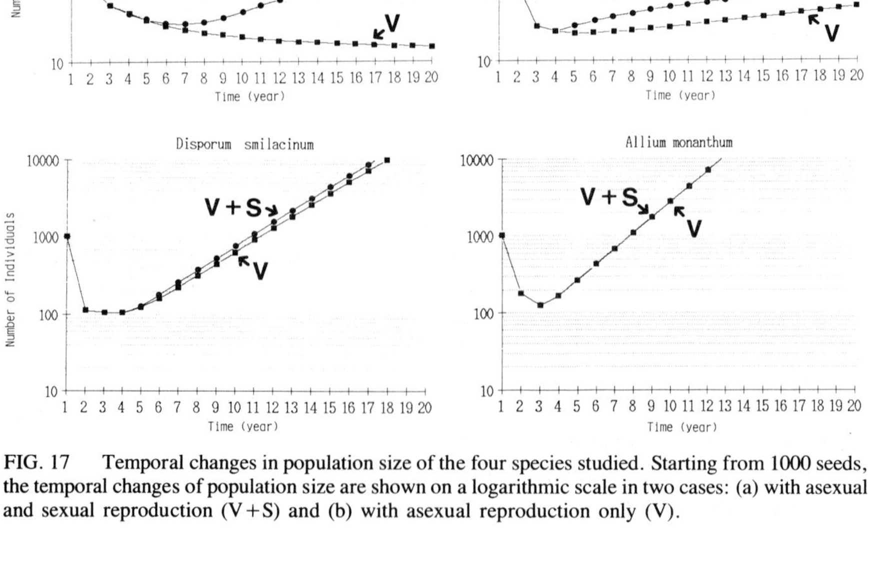

FIG. 17 Temporal changes in population size of the four species studied. Starting from 1000 seeds, the temporal changes of population size are shown on a logarithmic scale in two cases: (a) with asexual and sexual reproduction (V+S) and (b) with asexual reproduction only (V).

logarithmic scale in Fig. 17. Here we computed two cases: one with both sexual and asexual reproduction, and another without sexual reproduction, i.e., we use a modified Lefkovitch matrix with $a_{SE,j} = 0$ for all j. In the figure "V+S" indicates that both vegetative and sexual reproduction occur simultaneously, and "V" means only vegetative reproduction occurs. In *Erythronium japonicum*, a rapid decline in the number of individuals occurs when there is no sexual reproduction. On the other hand, sexual reproduction does not seem to play an important role in the offspring recruitment in the three pseudo-annual species studied, and it is virtually non-existent in *Allium monanthum*. Vegetative reproduction is very effective in pseudo-annual species; however, under field conditions various but as yet unknown mortality factors operate, and thus fluctuations of the local populations occur, depending upon the varying levels ("K") of environmental constraints.

IV. Conclusions

The complex life-history strategies and demographic patterns in the temperate woodland plants presented here may be related both to the environmental constraints prevailing in the temperate deciduous hardwood forests of the northern hemisphere, and also to changes in these factors which have occurred in the evolutionary past.

The Lefkovitch matrix is very useful for demographic analysis in plants with size-dependent reproduction. The sensitivity analysis proposed by Caswell (1978) is helpful in identifying the most important stage-class in the life history of a given species. On the other hand, the flow matrix and flow diagrams used in our study show the differences in modes of offspring recruitment. The advantage of using flow diagrams is that one can assess almost immediately the most important developmental stage and flow of individuals in life-history schedules of a given species. Indeed, we have clearly demonstrated the differences in dependency on sexual and asexual reproduction in three pseudo-annual woodland herbs.

The unique life-history strategy represented by pseudo-annuals and found among various temperate woodland plants is undoubtedly the most specialized. The vegetative propagules found in such pseudo-annuals have obviously been derived from the originally underground stock organs in response to a decrease in the efficiency of sexual offspring recruitment. Strong interference, limited resource availability, and specialized pollinators in the woodland environments may possibly be responsible for the differentiation of such complex and unique life histories.

In addition to "pseudo-annuals", some other types of life-history strategies were found in herbs forming part of the woodland community (Kawano, 1985). The first group represents polycarpic perennials like *Erythronium japonicum*, described in detail in this chapter, in which no asexual reproduction occurs except in very rare cases (Kawano, 1982), and offspring for new generations are recruited by seeds alone (Kawano *et al.*, 1982). This type was also found in numerous other woodland herbs, e.g., *Trillium kamtchaticum, T. apetalon, T. tschonoskii, T. smallii* (Ohara and Kawano, 1986), *T. grandiflorum, T. erectum, T. undulatum,* and *T. sessile* (Kawano *et al.*, 1986); all these species are representative of stable, closed temperate woodland communities.

The second group comprises polycarpic perennials with remarkable vegetative reproduction, as found for example in *Erythronium americanum* (Kawano and Utech, unpubl. data), and several *Trillium* species belonging to the sessile-flowered group. Ohara and Utech (1986) recently discovered that the sessile-flowered *Trillium* species growing in the outer Coastal Plain region of North America depend to a large extent on vegetative reproduction as a means of offspring recruitment to maintain their local populations. Vegetative reproduction observed among these species appears to be compensating for the inefficiency of sexual reproduction under somewhat unstable environmental conditions (e.g. flooding). In spite of the fact that the separation of vegetative propagules, depending upon the degree of environmental instability, is rather irregular, it is an effective way of maintaining local populations because the vegetative propagules manifest low mortality rates, absence of dormancy, and the rapid development to the critical size capable of producing sexual reproductive organs. This situation is somewhat different from that in "pseudo-annuals", in which vegetative propagules separate from their mother plants every season.

The third group is formed by monocarpic perennials, represented by *Cardiocrinum cordatum*, in which vegetative reproduction through bulblet formation occurs in sexually mature individuals after flowering (Kawano, 1985 and unpubl.).

The results obtained in the present study contribute further to the picture of life-history evolution in temperate woodland herbs emerging from our previous studies (Kawano, 1970, 1975, 1985). They demonstrate that the study of the comparative biology of the life-history characteristics of a species group or congeners which are members of the same community could be one of the most useful approaches in evolutionary biology.

References

Caswell, H. (1978). A general formula for the sensitivity of population growth rate to changes in life history parameters. *Theor. Popul. Biol.* **14**, 215–320.

Caswell, H. (1982). Stable population structure and reproductive value for populations with complex life cycles. *Ecology* **63**, 1223–1231.

Hartshorn, G. S. (1975). A matrix model of tree population dynamics. *In* "Tropical Ecological Systems" (F. B. Golley and E. Medina, eds), pp. 41–51. Springer, New York.

Kawano, S. (1970). Species problems viewed from productive and reproductive biology. I. Ecological life histories of some representative members associated with temperate deciduous forests in Japan. *J. College Liberal Arts, Toyama Univ.* **3**, 181–213.

Kawano. S. (1975). The productive and reproductive biology of flowering plants. II. The concept of life history strategy in plants. *J. College Liberal Arts, Toyama Univ. Nat. Sci.* **8**, 51–86.

Kawano. S. (1982). On the abnormal vegetative growth and reproduction in *Erythronium japonicum* Decne. (Liliaceae). *J. Phytogeogr. Taxon.* **30**, 53–55.

Kawano, S. (1984). Population biology and demographic genetics of some liliaceous species. *Korean J. Plant Taxon.* **14**, 43–57.

Kawano, S. (1985). Life history characteristics of temperate woodland plants in Japan. *In* "The Population Structure of Vegetation" (J. White, ed.), pp. 515–549. W. Junk, Dordrecht.

Kawano, S. and Nagai, Y. (1975). The productive and reproductive biology of flowering plants. I. Life history strategies of three *Allium* species in Japan. *Bot. Mag. Tokyo* **88**, 281–318.

Kawano, S., Ihara, M. and Suzuki, M. (1968). Biosystematic studies on *Maianthemum* (Liliaceae-Polygonatae). II. Geography and ecological life history. *Jpn. J. Bot.* **20**, 35–65.

Kawano, S., Hiratsuka, A. and Hayashi, K. (1982). The productive and reproductive biology of flowering plants. V. Life history characteristics and survivorship of *Erythronium japonicum. Oikos* **38**, 129–149.

Kawano, S., Ohara, M. and Utech, F. H. (1986). Life history studies on the genus *Trillium* (Liliaceae). II. Reproductive biology and survivorship of four eastern North American species. *Plant Species Biol.* **1**, 47–58.

Lefkovitch, L. P. (1965). The study of population growth in organisms grouped by stages. *Biometrics* **21**, 1–18.

Leslie, P. H. (1945). On the use of matrices in population mathematics. *Biometrika* **33**, 183–213.

Leslie, P. H. (1948). Some further notes on the use of matrices in population dynamics. *Biometrika* **35**, 213–245.

Nagai, Y. (1972). Reproductive biology of *Allium monanthum* Maxim. *J. Geobot.* **20** (in Japanese).

Ohara, M. and Kawano, S. (1986). Life history studies on the genus *Trillium* (Liliaceae). I. Reproductive biology of four Japanese species. *Plant Species Biol.* **1**, 35–46.

Ohara, M. and Utech, F. H. (1986). Life history studies on the genus *Trillium* (Liliaceae). III. Reproductive biology of six sessile-flowered species occurring in the southeastern United States with special reference to vegetative reproduction. *Plant Species Biol.* **1**, 35–46.

Salisbury, E. J. (1942). "The Reproductive Capacity of Plants". Bell, London.

Werner, P. A. and Caswell, H. (1977). Population growth rates and age versus stage-distribution models for teasel (*Dipsacus sylvestris* Huds.). *Ecology* **58**, 1103–1111.

9 Plant Evolution and Speciation on Oceanic Islands

DANIEL J. CRAWFORD, RICHARD WITKUS and TOD F. STUESSY

Department of Botany, The Ohio State University, Columbus, Ohio, U.S.A.

I. Introduction

It has long been recognized that congeneric species of plants on oceanic islands may differ by more conspicuous morphological features than distinguish congeners on continents. In some instances early taxonomic treatments recognized several genera to accommodate the variability; two such examples are *Dendroseris* on the Juan Fernandez Islands (Stuessy *et al.*, 1984a) and *Tetramolopium* on the Hawaiian Islands (Lowrey, 1981). The 11 species of *Tetramolopium* include such diverse life forms as succulent-leaved, cushion-forming plants and cespitose alpine shrubs. Ganders and Nagata (1984) comment that greater morphological variability occurs in the 27 taxa of *Bidens* in the Hawaiian archipelago than is found in the other 200 or so species occurring on five continents. Perhaps the most striking example from plants of morphological diversity on islands is the so-called silversword alliance comprising three genera in the Compositae (Carr and Kyhos, 1981; Carr, 1985a,b). This natural, doubtlessly monophyletic assemblage includes life forms such as cushion plants, semi-shrubs, shrubs, trees, monocarpic rosette plants and lianas (Carr, 1985b).

The morphological variation displayed by congeners on islands is often paralleled by the diversity of habitats in which they grow. The 11 species of *Tetramolopium* on the Hawaiian archipelago occur in a wide range of habitats from sea level to above 3000 m (Lowrey, 1981). Plants of the genus *Bidens* (19 species) occur on the Hawaiian Islands from sea level to over 2200 m and grow

Differentiation Patterns in Higher Plants
ISBN: 0-12-709425-3

in semi-deserts, rain forests and montane bogs (Ganders and Nagata, 1984). Species in the silversword alliance occur in a vast array of habitats in the Hawaiian Islands from alpine deserts to wet forests and bogs; annual precipitation ranges from 0.4 m to 16 m and the altitudinal distribution is from near sea level to 3750 m (Carr, 1985b). These examples represent some of the most extreme and thoroughly studied groups; for many additional examples, see Carlquist (1974).

Another noteworthy aspect of plants on oceanic islands is that population sizes are usually quite small compared to relatives in continental areas, or at least it appears that sizes are small at some time during the history of a population.

II. Ecological Diversity and Differentiation Patterns

A. *Habitats on oceanic islands*

Any understanding of the reasons for the great diversity of morphological features found among congeneric species on islands must rest in part on considerations of island habitats. Why, for example, should there be a greater range of morphological variability in the Hawaiian *Bidens* than in the rest of the genus as indicated by Ganders and Nagata (1984)? We will mention some of the attributes of island ecosystems which may be important for explaining the occurrence of congeners with different morphologies and which occur in very different habitats.

One aspect of many oceanic islands is their recent origin (see Stuessy *et al.*, 1984b; Hickman and Lipps, 1985, for dates of the Juan Fernandez and Galapagos Islands, respectively) as compared to many continental areas. Habitats on oceanic islands have been present for shorter periods of time than on continents. Thus little to no habitat pre-conditioning by other plants or animals takes place on islands before the invasion of new plant species, a situation similar to continental areas where some major force, such as glaciation, has completely wiped out any evidence of pre-existing communities.

Another aspect of oceanic islands is the diversity of habitats available in a small geographic region. In island archipelagos, the individual islands often form sequentially, resulting in islands of different ages. A prime example of this is the Hawaiian archipelago, where there has been a sequential development of islands resulting from an immobile hot spot (Wilson, 1963; McDougall, 1971; Morgan, 1972). This provides new habitats relatively near source propagules on previously formed islands. In addition, the mountains produced by volcanoes on islands can result in considerable variation in moisture and temperature. Erosion and active volcanoes can also create diverse habitats over short distances, thus reducing population sizes and producing new, open habitats very quickly.

Lastly, the isolated nature of oceanic islands contrasts sharply with most continental situations because of the rarity of dispersal of propagules to them. Propagules can become established in a new and open habitat without being swamped out by propagules of better adapted species, as would occur in continental areas.

B. Adaptive radiation on insular archipelagos

The process of adaptive radiation has been invoked to explain the diversity of species on oceanic islands. Adaptive radiation involves the adaptation of populations to a diversity of habitats (or niches) following dispersal to open niches. A general definition of the process given by Stebbins (1977) is, "Individuals entering new regions will found subpopulations which will differ in various adaptive characteristics from the ancestral population and from each other."

As pointed out by Carlquist (1974, p. 97), the concept of adaptive radiation is usually not explicitly defined, but is applied to particular instances and the examples become the definitions. The following basic ideas on application of the term adaptive radiation to evolution and speciation on islands are taken almost entirely from Carlquist (1974). These considerations will serve as a framework for the ensuing discussion of speciation on oceanic islands.

The term adaptive radiation implies that a group enters into a variety of different habitats. Furthermore, these diverse habitats occur in a well-defined geographical area such as an archipelago. Any plant group exhibiting adaptive radiation should have representative species in many, but probably not all, of these habitats. That is, diversification by a plant group into a wide variety of habitats is considered to be an important criterion of adaptive radiation.

The study of adaptive radiation in a particular group on an island or archipelago assumes that the group is monophyletic, as a result of a single introduction or invasion. This seems self-evident, but it is not always a simple matter to establish. If a genus is very distinct and endemic to a particular island or archipelago, then one can assume monophyly with confidence. This may not be the case, however, with a group of endemic species in a genus having representatives elsewhere. Establishing monophyly may best be accomplished by using characters not important in adaptation to the diverse habitats. Molecular attributes such as isozymes, chloroplast DNA, or ribosomal DNA are probably of greatest utility for ascertaining the monophyletic nature of a group.

The concept of adaptive radiation has been applied at a variety of taxonomic levels. In a broad sense, it has been invoked for explaining aspects of the origin and evolution of the angiosperms. For insular situations, the most meaningful and useful taxonomic categories are the subgenus, genus, a group of closely related genera, or a subfamily. These levels are also amenable to experimental studies in ways not possible with higher taxonomic categories.

It must be emphasized that the development of morphological and ecological diversity within a single species is not normally viewed as adaptive radiation. Rather, it may be considered an "incipient form of it" (Carlquist, 1974, p. 102) and the process results in the formation of ecological races. This leads to another aspect of adaptive radiation on islands, namely that discontinuities develop in the group so that distinct morphological types with well-defined ecological preferences can be recognized. These different forms are normally recognized taxonomically as subspecies or species. A morphologically variable taxon occurring in a wide range of habitats would not be considered an example of adaptive radiation if the morphological forms intergrade and occur continuously over a variety of habitats. As Carlquist (1974) suggests, the ability to designate "good" taxonomic species is often considered a prerequisite for viewing something as an example of adaptive radiation. Beyond this, Carlquist (1974, pp. 106–107) further suggests that adaptive radiation results in "strikingly different, often exceptional" products.

One implication of adaptive radiation is that the basic process of diversification has occurred quite recently. In absolute terms, this usually means several thousand to several million years. Because the ages of oceanic islands can often be determined with reasonable accuracy, it is possible to set the oldest age for the taxa endemic to given islands. Certain precautions must be taken when doing this, however, because geological and biological factors may allow for a taxon that originated on an older island of an archipelago to disperse to its present area of occurrence. Carlquist (1974, pp. 104–105) makes the important point that, for many groups, recent origin is the best explanation for their occupying a wide variety of niches on islands. Lowered predation and competition may permit long-term persistence and thus account for the presence of members of a group in many different habitats, but recentness of diversification seems the overriding consideration.

If adaptive radiation occurs on islands as outlined above, then certain predictions should hold with regard to divergence between species in certain features. Available data will be reviewed and discussed in the context of adaptive radiation.

C. *Interfertility of congeners on islands*

The genera *Bidens* and *Tetramolopium*, plus the three genera making up the silversword alliance, have been used to illustrate the morphological and ecological diversity among congeners on oceanic islands. These three groups are also among the few insular taxa in which artificial interspecific crosses have been carried out and in which naturally occurring hybrids have been studied in detail for fertility. Lowrey (1981) showed that all extant taxa of *Tetramolopium* are highly interfertile through the F_3 generation. There is no evident loss of fertility or viability, with chromosome pairing and segregation normal in meiosis in the hybrids. The studies of Gillett (1972), Gillett and Lim (1970) and Ganders and Nagata (1984) in *Bidens* demonstrated high interfertility among all the taxa tested, with the majority of recognized taxa included in these studies. The elegant cytogenetic studies of Carr and Kyhos (1981, 1986) demonstrated that in the silversword alliance chromosomal repatterning (aneuploidy and reciprocal translocations) lowers the fertility of some hybrid combinations. Given, however, the pronounced morphological diversity in the group (with three genera recognized), the interfertility of most hybrid combinations is quite remarkable. Carr and Kyhos (1981) commented that the situation in the silversword alliance, i.e., chromosomal repatterning and reduced fertility of interspecific hybrids, is exceptional if not unique among plant groups in Hawaii.

Higher interfertility is to be expected if speciation has occurred as envisioned by adaptive radiation. It appears that in many continental plant groups, particularly perennials, post-mating reproductive isolating barriers develop with time following initial geographical and/or ecological divergence (i.e., adaptive divergence). If divergence has been recent in insular groups, then little post-mating isolation would be expected.

Although congeneric species on oceanic islands may lack gametic or post-zygotic isolating mechanisms, natural hybridization in several well-studied groups (*Bidens, Tetramolopium, Lipochaeta*, the silverswords, and *Dendroseris*) appears to be rare or non-existent (Ganders and Nagata, 1984; Lowrey, 1981; Gardner, 1979; Carr, 1985b; Stuessy *et al.*, 1984a). Natural interspecific hybridization does occur in certain groups, but its importance as an evolutionary factor on islands has not been documented, and in certain instances (as with Hawaiian *Bidens*), natural hybrids may not be nearly as common as earlier estimates indicated (see Ganders and Nagata, 1984). There appears little doubt that geographical and ecological separation prevent hybridization from being much more common on islands.

D. Genetic divergence among congeners on oceanic islands

1. Molecular data. In contrast to the situation with animals on islands, where enzyme electrophoresis and other approaches have been employed for measuring genetic divergence between taxa in molecular attributes, few such studies appear to be available for plants. Lowrey and Crawford (1985) examined allozymes in the extant taxa of Hawaiian *Tetramolopium*, and found essentially no divergence among them. Populations of seven species exhibited a mean genetic identity of 0.95; this is comparable to values found for conspecific populations. That is, despite the morphological and ecological diversity among the taxa (as described above), populations of different taxa exhibit the same genetic identities as those of the same taxon. In addition, populations of Hawaiian *Tetramolopium* contain very little variation in genes specifying soluble enzymes, with the proportion of polymorphic loci in the average population being less than 4% and observed heterozygosity less than 1% for all plants. Digestion of chloroplast DNAs from nearly all extant taxa of Hawaiian *Tetramolopium* with six restriction enzymes failed to reveal any differences among the taxa (Lowrey, unpubl.).

Crawford *et al.* (1987) examined six species from two of the three recognized subgenera of the genus *Dendroseris* (Compositae: Lactuceae) which is endemic to the Juan Fernandez Islands, Chile. Despite large morphological differences in floral and vegetative features among the subgenera, very little divergence was detected in populations of any species of these tetraploid ($n = 18$) plants. Thus, *Dendroseris* is similar to *Tetramolopium* in isozymes because of low levels of variation and lack of divergence among the taxa. Limited preliminary studies with restriction analysis of chloroplast DNA in five species of *Dendroseris* failed to reveal restriction site differences for three enzymes.

Helenurm and Ganders (1985) studied allozymes in 22 populations representing 15 taxa of Hawaiian *Bidens*. They found for pair-wise comparisons of all populations genetic identities ranging from 0.886 to 0.996, with a mean value of 0.957. Thus, the results for Hawaiian *Bidens* are similar to those for *Tetramolopium* and *Dendroseris* in the lack of divergence between taxa at isozyme loci. *Bidens* differs from the other genera, however, in that individual populations of these hexaploid plants harbour much higher levels of variation than the other two genera, with observed heterozygosity of over 10% (Helenurm and Ganders, 1985).

The patterns of divergence in morphological and molecular features among many insular plant species differ from the common situation in continental taxa. Species that are very different morphologically yet lack divergence in isozymes and chloroplast DNA are essentially unknown in continental taxa. On land masses, species exhibiting high genetic identities at genes specifying isozymes are usually annual plants which are very similar morphologically, and often consist of pairs of species thought to be related as progenitor and derivative (see Gottlieb, 1977, 1981, 1984a; Crawford, 1983, 1985). Most morphologically well-differentiated species exhibit genetic identities of *c*. 0.65 (although there is a considerable range of variation among all taxa examined) as opposed to values above 0.90 for most congeners on islands. Also, most plant species that have been subjected to restriction analysis of chloroplast DNA are distinguishable (see Palmer, 1985, and in press for excellent reviews of the topic).

The most plausible explanation for the non-concordance between molecular data on the one hand and morphological and ecological features on the other in island species is that the tempo or rate of change is much faster in the latter two features as opposed to the former. Morphological characters (at least some of them) and ecological preferences are important in adaptive evolution and speciation on islands. By contrast, this is not so for molecular features such as allozymes and mutations at restriction sites in chloroplast DNA, and thus there is little reason to believe that the rate of change in these molecules in insular taxa differs from plants on continental areas. In short, results from molecular data are what would be expected if divergence were recent on islands.

2. *Genetic basis of morphological differences.* Certain basic questions arise after considering the evolution of such diversity among congeneric species in different genera on oceanic islands. One question concerns why this happens in insular situations but is extremely rare elsewhere. Another question is whether the differences are more apparent than real; it may be that rather small genetic differences underlie many of the rather considerable differences in morphology and/or ecological preferences.

There is no reason to believe that the founder plants from which insular species develop are any more variable genetically than other members of the genus growing elsewhere. If anything, insular species originate from a small number of founders which carry a smaller amount of the genetic variability than occurs in the ancestral species as a whole. Also, given the presumed recent origin of many insular species, it seems doubtful that mutation could account for the generation of the variability seen on islands.

The wide variety of niches available within small geographical areas on oceanic islands no doubt is an important factor for the presence of the diversity among congeneric species. That is, ecological opportunities are available and variants generated by recombination are not constrained by the physical environment or other organisms as they might be in other situations. Thus, it may be that many groups have the potential to diverge and diversify but it is only when the opportunity is available that the potential is achieved. Rick (1983) has documented the presence of mutant genes which obviously reduce vigour in tomatoes on the Galapagos Islands. He indicates that the array of mutant genes is "unique among tomato species", and attributes their presence to drift and reduced selection on the islands.

The genetic basis of morphological differences between species is poorly understood, although it is the topic of renewed interest among plant evolutionists (see Bachmann, 1983; Hilu, 1983; and Gottlieb, 1984b for recent reviews). The genetic bases for the differences among island congeners are likewise undetermined. It is possible that a number of the most conspicuous differences *may have* a relatively simple genetic basis. For example, in *Bidens* and *Tetramolopium* on the Hawaiian Islands, congeners in one or both of the genera are separable by such features as achenes winged or wingless; achenes awned or awnless (or uniserate or biserate pappus), achenes pubescent or glabrous, capitula erect or nodding, growth habitat erect or prostrate, large solitary capitulum or several aggregated capitula, and colour of corolla of disk florets (or ray florets). All of these kinds of differences have been shown to be governed by one to several genes or major segregating units in particular plant taxa (see list in Gottlieb, 1984b). It is not being argued that similar features are necessarily inherited in the same manner in *Bidens, Tetramolopium*, or other congeners on islands. Rather, the point is that it is *possible* that certain rather obvious morphological features have a simple genetic basis, and that a combination of several of these features may cause two species to look strikingly different. In short, the differences may be more apparent morphologically than they are real genetically. We do not intend to argue that all the differences among congeners on oceanic islands have a simple genetic basis, but suggest rather than extensive genetic differences may not be necessary to produce what appears to be very distinct species.

Coyne and Lande (1985) have challenged certain of the conclusions made by Gottlieb (1984b), with Gottlieb (1985) then responding to these criticisms. One of the points criticized is that pleiotropic effects associated with differences based on single genes of large effects might be deleterious. Coyne and Lande also comment that deleterious effects associated with single gene differences not selected against in domesticated plants, might not persist under natural conditions. These two factors could be of some importance in adaptive radiation on islands because reduced competition on islands may minimize negative pleiotropic effects and allow variants based on genetic differences at one or a few genes to become established (Rick, 1983). After establishment, adaptive "fine tuning" based on selection at other loci could occur.

Inferring genetic differences associated with ecological preferences displayed by congeneric species on oceanic islands is difficult at best. It is known that divergence can occur within populations over a very short period of time or over a short distance in response to various edaphic conditions (see McNeilly, 1968; Wu *et al.*, 1975; Wu and Antonovics, 1976; and Snaydon and Davies, 1982 as examples). In addition, several studies of the genetic bases of such features as heavy metal tolerance have been carried out, and the results sometimes suggest a relatively simple inheritance of these attributes (see Urquhart, 1971; Gartiside and McNeilly, 1974; Macnair, 1983 as examples). There is also evidence that drought tolerance can be under rather simple genetic control (Teeri, 1978).

The different ecological preferences of congeners on oceanic islands, while at first seeming unusually diverse for plants, are matched by the ecotypes and subspecies found in continental species. The point is that while the range of ecological diversity of congeneric species or different subspecies in insular situations may appear initially to be unique, the fact that the different habitats are telescoped in space on the islands only makes this appear to be the case.

III. Models of Plant Evolution and Speciation on Islands

Having reviewed the data for island plants, we propose some simple models to account for adaptive radiation and speciation on islands. The models are given only as a possible means by which island speciation via adaptive radiation may occur, and it is recognized that certain simplifying assumptions, which must be made in all models, are probably not met in all instances.

Wright's (1932) concept of *adaptive topography* will be employed as a means of illustrating several aspects of adaptive radiation. The adaptive topography is a device for illustrating mean population fitness values over a multi-dimensional space in which each axis represents the frequency of a particular gene. The adaptive topography concept is an integral part of Wright's (1932, 1977) shifting balance theory of evolution.

Wright conceived of the shifting balance theory of evolution (see Wright, 1970, 1980 for general discussions) as a means of explaining adaptive transformation within a species, but noted it may also be used for the adaptive divergence of populations during speciation (Wright, 1970, 1977, 1982). The theory is based on several premises dealing with genetic organization within organisms and with population structure, these being the multiple factor theory of quantitative variation, universal pleiotropy, multiple selective peaks within a given environment, and the existence of multiple, partially isolated demes.

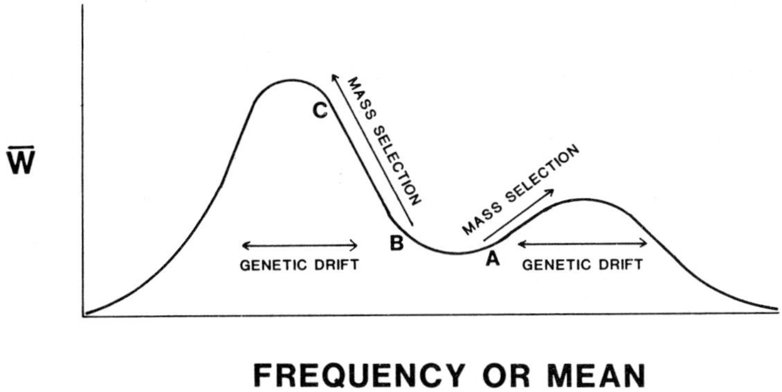

FIG. 1 Adaptive landscape and shifting balance theory of Wright (1932, 1977). The abscissa represents gene frequency (after Wright, 1977), or values of a character mean (after Simpson, 1953). The ordinate represents mean fitness. See text for further explanation.

Wright's shifting balance theory consists of three processes (Fig. 1). A deme within a population (point A) is located on a particular selective peak and driven towards the apex by *mass selection*. One force counterbalancing selection is *random genetic drift*, which continually diverts the population away from the apex. In a small enough deme, the action of drift can overcome selection and knock the population onto a saddle between two peaks (point B). The deme now is brought under the influence of a new selective peak where steepness of the sides is a measure of selective intensity. The deme climbs the new peak by mass selection until the forces of drift and selection are again counterbalanced (point C). Being located on a higher selective peak indicates the deme is better adapted to the surrounding environment. Immigration from this deme to other demes can change gene frequencies in the latter to the point where they will also cross the saddle and climb to the higher selective peak. This phase of *interdemic selection* (not shown) continues until all demes within a population are located on the new, higher peak.

To apply the shifting balance theory to island speciation, some modifications and elaborations of the premises are made to account for the situation on islands and to derive a model for adaptive radiation. Although the shifting balance theory is based on multiple factor theory of quantitative variation, adaptation by major mutations could also apply (Wright, 1977). However, Wright feels that mutations with major effects are usually deleterious and strongly selected against. If these deleterious side-effects can be removed by suitable combinations of modifier genes, then the model applies. Templeton (1982) describes a genetic architecture which consists of one or a few major loci with epistatic modifiers controlling isolating barriers among species. This type of genetic system is more effective in the shifting balance theory since drift affecting a few major genes is more likely to lead to an adaptive peak shift than when many segregating units are involved (Templeton, 1982). A genetic system of this nature is in line with data presented by Bachmann

(1983), Hilu (1983) and Gottlieb (1984b) on the apparent simple genetic basis of many characters in plants. Although these apparent single-gene systems may have a more complex basis (Coyne and Lande, 1985), they probably represent a system of one or a few genes with major effects in combination with modifier loci (Gottlieb, 1984b, 1985). The same may hold for ecological adaptations in plants (Urquhart, 1971; Gartiside and McNeilly, 1974; Macnair, 1983). Thus, whether the morphological and ecological differences seen in island plants represent simple genetic changes or genes with major effects having modifiers, both are applicable to Wright's model.

Several modifications or elaborations can be made to the idea of *multiple selective peaks*. Multiple selective peaks in the adaptive landscape exist because of the myriad of ways in which genotypes respond to a particular habitat. The habitat itself represents part of the multi-dimensional combination of factors representing abiotic and biotic components which comprise the niche (Hutchinson, 1968). This inexorably links the adaptive landscape to the concept of the niche, with a change in niche space resulting in a change in adaptive topography.

Simpson (1953) viewed the fitness of phenotypes rather than genotypes as a series of peaks on an adaptive landscape. This concept has been formalized by Lande (1976). As a consequence of this view, a peak shift in a population represents a change in mean population phenotype as well as mean population genotype.

The works of Simpson (1953), Wright (1970, 1977, 1980, 1982), Lande (1980), and Carson and Templeton (1984) indicate that major evolutionary change may occur if a species can escape the confines of its present niche and exploit new directions and areas of an adaptive landscape. This change in the adaptive landscape is brought about by a colony arriving in a new territory where open niches are available for exploitation, or by pre-adaptation to a drastically altered environment, or by increased specialization which opens up a new way of life to the species (Wright, 1970, 1977, 1982). These alternatives are summed up by Wright (1970, 1977) as *ecological opportunity*. We believe one way to view this is an expanded adaptive topography. For a particular species, the expanded landscape consists of the actual selective peaks a species now exists on, as well as *all* possible, favourable peaks for a given niche. These would include series of peaks which do not exist in the present niche because of unavailable niche space from competition with other species, as well as peaks that remain unoccupied because of local extinction or because the right combination of genes has not been realized in the species to allow it to occupy those peaks. The expanded topography would show a series of peaks separated by relatively shallow saddles, and these in turn are separated from other such series by deeper saddles. The only way to cross these deeper saddles is by extreme genetic changes in a population, such as those afforded by founder events.

Finally, it should be noted that even within a particular niche, the multiple selective peaks are not static in their height nor steepness of slope. Density-dependent selection can alter peaks, providing another complicating factor in the adaptive topography (Rosenzweig, 1978). For our purposes, we shall not incorporate this aspect of the adaptive topography into our models. In many island situations, population sizes may often be small enough in plants to avoid the problem of density-dependent selection.

A basic model for speciation on islands is presented in Fig. 2(a). A founder event occurs, and in plants this may represent one to several seeds. A single-seed founder carries very little genetic variation as compared to a single gravid female for animal founder events, since the seed contains only one zygote, while a gravid

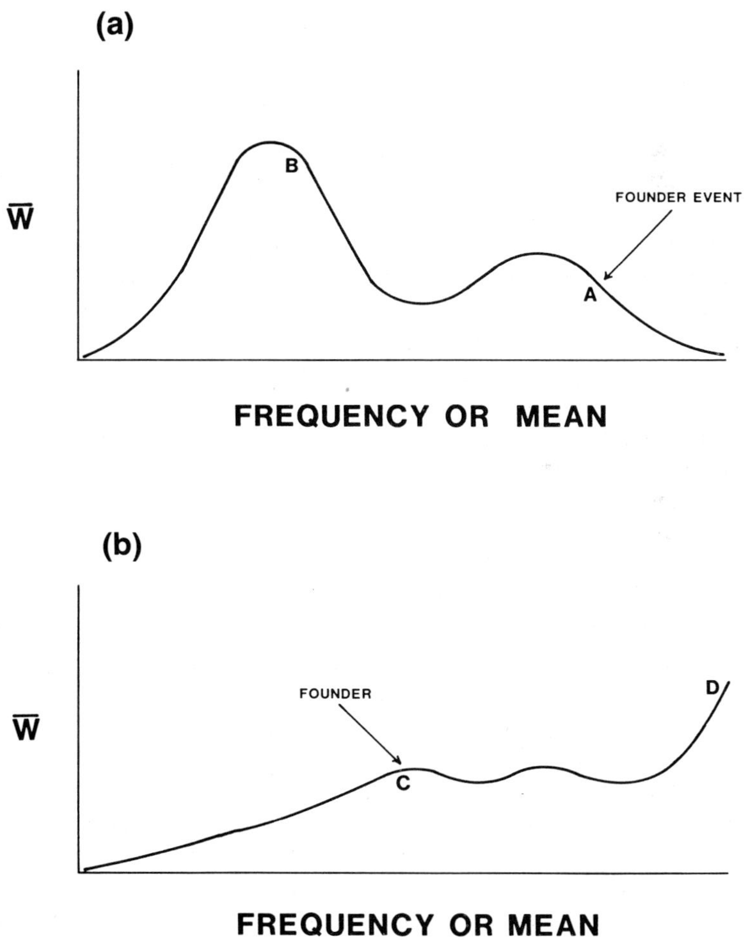

FIG. 2 Application of shifting balance theory to island speciation. (a) Adaptive landscape for a given island niche. (b) Adaptive landscape for a different island niche. Axes the same as in Fig. 1. See text for further explanation.

female carries many. The initial population starts out on a selective peak in an environment (niche) which is generally different from that of the ancestral population. Through time, drift and selection drive the founder population up to the highest selective peak for that niche (point B). Dispersal of individuals from this population to new areas on the island or into the same niche on other islands initiates populations which eventually will shift to the same selective peak. The small size of island populations tends to maintain drift as an important factor, allowing populations to explore the adaptive landscape. But once on the highest selective peak, it becomes more difficult for drift to knock the population off the peak. From this model, we would expect island populations of one particular species to be phenotypically, ecologically and genetically similar. Data for the plant groups studied indicate that this is the case.

Once a species becomes established on an island, dispersal into new niches alters the selective topography and leads to divergence of populations (Fig. 2(b)). Populations in new niches will climb new selective peaks (point D), becoming reproductively isolated from the island progenitor species due to differences in habitat. The new selective peaks also represent phenotypic divergence (Simpson, 1953), and so species become morphologically divergent as well. If the number of genetic changes needed to produce these phenotypic and ecological changes is few, then the species are expected to be genetically similar, with speciation following the genetic transilience of Templeton (1981). Multiply this process by the number of new niches a species finds on an island group (Wright's ecological opportunity), and adaptive radiation can occur quickly, producing ecologically and morphologically diverse species in very little time.

We may also use the expanded adaptive topography as a way of viewing adaptive radiation (Fig. 3). The steps are the same as in Fig. 2, except founders carry populations across very deep saddles into areas of the adaptive topography which species cannot explore in continental areas because of gaps or closures in niche space (point A). This could result in more extreme divergence among the resultant species with large reproductive barriers built up among them, as may be the case in the silversword alliance.

Given the proper conditions, we feel these simple models can explain adaptive radiation on islands. A word of caution is necessary in accepting these models. The conditions under which speciation via adaptive radiation can occur are not easily obtained. Founder events are rare, and many are likely to lead to extinction of the founder population. Once established, the founder event does not by itself induce speciation (Mayr, 1963; Carson and Templeton, 1984), but provides the opportunity for it to occur. Island situations appear to be most conductive to speciation via founder events as evidenced by the data presented in this review and elsewhere.

Once adaptive radiation is initiated, the condition for it to occur in a relatively short period of time and result in morphologically and ecologically divergent species, infers a genetic transilience (Templeton, 1980a,b, 1981). This is a rapid

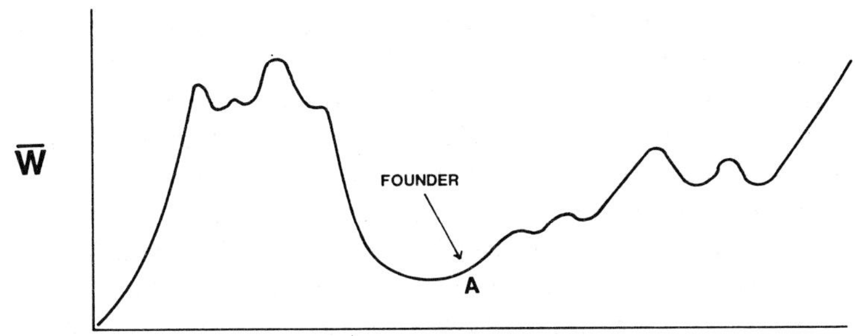

FIG. 3 Expanded adaptive landscape. The series of peaks on the left represent the portion of the adaptive landscape a species is able to explore within the confines of a particular niche and genetic system, and is analogous to Figs 1 and 2. The series of peaks on the right represent the areas on an adaptive landscape which a species may explore given an open habitat or proper gene combinations. Axes the same as previous figures. See text for further explanation.

change in a genetic system, induced by a founder event, leading to alteration of a few major alleles with reproductive isolation arising as a direct or pleiotropic consequence (Templeton, 1981). Again, the conditions for this sequence to occur are rare and stringent. But given the patterns seen for both plant and animal groups on islands, it appears to be the best explanation.

IV. Conclusion: Islands as Laboratories for the Study of Plant Evolution

Future research efforts on speciation in island plants should be modelled to some extent after the classic *Drosophila* studies on the Hawaiian Islands. An important initial aspect is ascertaining that a group is monophyletic and to produce a phylogeny for it. This goal is best accomplished by use of neutral characters such as isozymes or chloroplast DNA. Use of morphological features in constructing phylogenies for island plants may yield less than satisfactory results because of parallel evolution. Once a phylogeny has been produced it is possible to study and compare the most closely related species and to infer changes associated with speciation.

A basic problem with studying speciation is ascertaining those differences between species that were involved in speciation versus accumulated changes subsequent to divergence. As stated by Templeton (1981), the basic task is, "distinguishing between the genetics of speciation and the genetics of species differences." The more recent that speciation has occurred, the higher the feasibility of distinguishing between between the two. The recency of island habitats as compared to many continental areas assumes that insular species are of more recent derivation than most of their counterparts on mainlands. The lack of divergence in molecular characters in plant species on islands lends further support to the hypothesis of recent origin. Because the post-mating isolating mechanisms occurring between congeners on islands are very weak or nil, it is possible to produce interspecific hybrids. Inferences may then be made about the genetic bases of morphological features distinguishing species, and those associated with speciation. Since congeners on islands diverged into different habitats, F_1 hybrids and later segregation generations can provide evidence on morphological features correlated with adaptation to a particular habitat. Not only should morphological features of the synthetic hybrids be considered, but segregation for the ability to grow and reproduce in different habitats should also be examined. These data may be obtained from transplant and competition experiments between parental taxa as well as hybrids. Further experimentation may include the production of artificial intermediate habitats to test further the performance of parental taxa and their hybrids. More studies of the kind done by Robichaux (1984) are needed. He showed that the two sympatric species *Dubautia ciliolata* and *D. scabra* differ in their abilities to maintain high turgor pressures when water content of tissue decreases. Natural hybrids between the two species show intermediate properties. These kinds of studies address basic questions, and they are more feasible with plants than with animals.

The features of plants normally examined are associated with the mature sporophyte; this is natural because island species have been studied primarily by plant taxonomists. Other aspects of the organisms' life history may be as critical if not more critical for survival. Seed size and seedling establishment are important factors worthy of study (see Gross, 1984; Gross and Kromer, 1986; Wulff, 1986a,b). There is, for example, a ten-fold difference in the weights of seeds of different species of *Dendroseris* on the Juan Fernandez Islands (Crawford and Stuessy, unpubl.), yet the significance of these differences is not known.

It is puzzling that even though islands are considered to be ideal localities for studying plant evolution, very little modern work has been done on insular groups. While traditional taxonomic revisions are needed for given taxa as a first step, it seems that additional, more in-depth studies are rarely carried out. Rather, more revisions of additional groups are done. Plants have much to offer as experimental organisms and their potential should be exploited. In order for this to be accomplished, various kinds of ecologists, geneticists and persons interested in plant evolution *per se* must become involved. The study of plant speciation on islands promises to be a rewarding endeavour; one waiting to be realized with expanded efforts.

Acknowledgements

We appreciate the financial support of grants BSR-8306436, DEB 80-11055 and DEB 82-04073 from The National Science Foundation. We also appreciate very much the useful comments made by Kay Gross on the manuscript.

References

Bachmann, K. (1983). Evolutionary genetics and the genetic control of morphogenesis in flowering plants. *Evol. Biol.* **16**, 157–208.

Carlquist, S. (1974). "Island Biology". Columbia University Press, New York.

Carr, G. D. (1985a). Habitat variations in the Hawaiian Madiinae (Heliantheae) and its relevance to generic concepts in the Compositae. *Taxon* **34**, 22–25.

Carr, G. D. (1985b). Monograph of the Hawaiian Madiinae (Asteraceae): *Argyroxiphium, Dubautia* and *Wilkesia. Allertonia* **4**, 1–123.

Carr, G. D. and Kyhos, D. W. (1981). Adaptive radiation in the Hawaiian silversword alliance (Compositae: Madiinae). I. Cytogenetics of spontaneous hybrids. *Evolution* **35**, 543–556.

Carr, G. D. and Kyhos, D. W. (1986). Adaptive radiation in the Hawaiian silversword alliance (Compositae: Madiinae). II. Cytogenetics of artificial and natural hybrids. *Evolution* **40**, 959–976.

Carson, H. L. and Templeton, A. R. (1984). Genetic revolutions in relation to speciation phenomena: the founding of new populations. *Ann. Rev. Ecol. Syst.* **15**, 97–131.

Coyne, J. A. and Lande, R. (1985). The genetic basis of morphological differences in plants. *Am. Nat.* **126**, 141–145.

Crawford, D. J. (1983). Phylogenetic and systematic inferences from electrophoretic studies. *In* "Isozymes in Plant Genetics and Plant Breeding" (S. O. Tanksley and T. J. Orton, eds), Part A, pp. 257–287. Elsevier, Amsterdam.

Crawford, D. J. (1985). Electrophoretic data and plant speciation. *Syst. Bot.* **10**, 405–416.

Crawford, D. J., Stuessy, T. F. and Siva, M. O. (1987). Allozyme divergence and the evolution of *Dendroseris* (Compositae: Lactuceae) on the Juan Fernandez Islands. *Syst. Bot.* **12**, 435–443.

Ganders, F. R. and Nagata, K. M. (1984). The role of hybridization in the evolution of *Bidens* on the Hawaiian Islands. *In* "Plant Biosystematics" (W. F. Grant, ed.), pp. 179–184. Academic Press, Orlando, FL.

Gardner, . C. (1979). Revision of *Lipochaeta* (Compositae: Heliantheae) of the Hawaiian Islands. *Rhodora* **81**, 291–343.

Gartiside, D. W. and McNeilly, T. (1974). Genetic studies in heavy metal tolerant plants. I. Genetics of zinc tolerance in *Anthoxanthum odoratum. Heredity* **32**, 287–297.

Gillett, G. W. (1972). The role of hybridization in the evolution of the Hawaiian flora. *In* "Taxonomy, Phytogeography and Evolution" (D. H. Valentine, ed.), pp. 205–219. Academic Press, London.

Gillett, G. W. and Lim, E. K. S. (1970). An experimental study of the genus *Bidens* (Asteraceae) in the Hawaiian Islands. *Univ. Calif. Publ. Bot.* **56**, 1–63.

Gottlieb, L. D. (1977). Electrophoretic data and plant systematics. *Ann. Mo. Bot. Gard.* **64**, 161–180.

Gottlieb, L. D. (1981). Electrophoretic evidence and plant populations. *Progr. Phytochem.* **7**, 1–46.

Gottlieb, L. D. (1984a). Isozyme evidence and problem solving in plant systematics. *In* "Plant Biosystematics" (W. F. Grant, ed.), pp. 343–357. Academic Press, Orlando, FL.

Gottlieb, L. D. (1984b). Genetics and morphological evolution in plants. *Am. Nat.* **124**, 681–209.

Gottlieb, L. D. (1985). Reply to Coyne and Lande. *Am. Nat.* **126**, 146–150.

Gross, K. L. (1984). Effects of seed size and growth form on seedling establishment of six monocarpic perennials. *J. Ecol.* **72**, 369–387.

Gross, K. L. and Kromer, M. (1986). Seed weight effects on growth and reproduction in *Oenothera biennis. Bull. Torrey Bot. Club.* **113**, 252–258.

Helenurm, K. and Ganders, F. R. (1985). Adaptive radiation and genetic differentiation in Hawaiian *Bidens*. *Evolution* **39**, 753–765.

Hickman, C. S. and Lipps, J. H. (1985). Geologic youth of Galapagos Islands confirmed by marine stratigraphy and paleontology. *Science* **227**, 1578–1580.

Hilu, K. W. (1983). The role of single gene mutations in the evolution of flowering plants. *Evol. Biol.* **16**, 97–128.

Hutchinson, G. E. (1968). When are species necessary? *In* "Population Biology and Evolution" (R. C. Lewontin, ed.), pp. 177–186. Syracuse University Press, Syracuse. NY.

Lande, R. (1976). Natural selection and random genetic drift in phenotypic evolution. *Evolution* **30**, 314–334.

Lande, R. (1980). Genetic variation and phenotypic evolution during allopatric speciation. *Am. Nat.* **116**, 463–479.

Lowrey, T. K. (1981). A biosystematic study of Hawaiian *Tetramolopium* (Compositae: Asteraceae). Ph.D. dissertation, University of California, Berkeley, CA.

Lowrey, T. K. and Crawford, D. J. (1985). Allozyme divergence and evolution in *Tetramolopium* (Compositae: Asteraceae) on the Hawaiian Islands. *Syst. Bot.* **10**, 64–72.

MacNair, M. R. (1983). The genetic control of copper tolerance in the yellow monkey flower, *Mimulus guttatus*. *Heredity* **50**, 283–293.

Mayr, E. (1963). "Animal Species and Evolution". Belknap Press, Cambridge, Mass.

McDougall, I. (1971). Volcanic island chains and sea floor spreading. *Nature Phys. Sci.* **231**, 141–144.

McNeilly, T. (1968). Evolution in closely adjacent plant populations. III. *Agrostis tenuis* on a small copper mine. *Heredity* **23**, 99–118.

Morgan, W. J. (1972). Deep mantle convection plumes and plate motions. *Bull. Am. Ass. Petr. Geol.* **56**, 203–213.

Palmer, J. D. (1985). Chloroplast DNA and molecular phylogeny. *Biol. Essays* **2**, 263–267

Palmer, J. D. (In press). Phylogenetic analysis of chloroplast DNA variation. *Ann. Mo. Bot. Gard.*

Rick, C. M. (1983). Genetic variation and evolution of Galapagos tomatoes. *In* "Patterns of Evolution in Galapagos Organisms" (R. I. Bowman, M. Bersen and A. E. Leviton, eds), pp. 97–106. Pacific Division AAAS, San Francisco.

Robichaux, R. H. (1984). Variation in the tissue water relations of two sympatric Hawaiian *Dubautia* species and their natural hybrid. *Oecologia* **65**, 75–81.

Rosenzweig, M. L. (1978). Competitive speciation. *Biol. J. Linn. Soc.* **10**, 275–289.

Simpson, G. G. (1953). "The Major Features of Evolution". Columbia University Press, New York.

Snaydon, R. W. and Davies, T. M. (1982). Rapid divergence of plant populations in response to recent changes in soil conditions. *Evolution* **36**, 289–297.

Stebbins, G. L. (1977). "Processes of Organic Evolution", 3rd Ed. Prentice-Hall, Englewood Cliffs, NJ.

Stuessy, T. F., Sanders, R. W. and Silva, M. O. (1984a). Phytogeography and evolution of the Juan Fernandez Islands: a progress report. *In* "Pacific Basin Biogeography" (P. A. Raven, F. Radonsky and S. Sohmer, eds), pp. 55–69. Stanford University Press, Stanford, CA.

Stuessy, T. F., Sutter, J. F., Sanders, R. W. and Silva, M. O. (1984b). Botanical and geological significance of potassium–argon dates from the Juan Fernandez Islands. *Science* **25**, 49–51.

Teeri, J. (1978). Environmental and genetic control of phenotypic adaptation to drought in *Potentilla glandulosa* Lindl. *Oecologia* **37**, 29–39.

Templeton, A. R. (1980a). Modes of speciation and inferences based on genetic distances. *Evolution* **34**, 719–729.

Templeton, A. R. (1980b). The theory of speciation via the founder principle. *Genetics* **94**, 1011–1038.

Templeton, A. R. (1981). Mechanisms of speciation — a population genetic approach. *Ann. Rev. Ecol. Syst.* **12**, 23–48.

Templeton, A. R. (1982). Genetic architectures of speciation. *In* "Mechanisms of Speciation" (C. Barigozzi, ed.), pp. 105–121. C. Barigozzi, Alan R. Liss, New York.

Urquhart, C. (1971). Genetics of lead tolerance in *Festuca ovina. Heredity* **26**, 19–33.

Wilson, J. T. (1963). A possible origin of the Hawaiian Islands. *Can. J. Phys.* **41**, 863–870.

Wright, S. (1932). The roles of mutation, inbreeding, crossbreeding, and selection in evolution. *Proc. VIth Int. Congr. Genet.* **1**, 356–366.

Wright, S. (1970). Random drift and the shifting balance theory of evolution. *In* "Mathematical Topics in Population Genetics", Biomathematics, Vol. 1 (K. Kojima, ed.), pp. 1–31. Springer, New York.

Wright, S. (1977). "Evolution and the Genetics of Population", Vol. 3: Experimental results and evolutionary deductions. University of Chicago Press, Chicago.

Wright, S. (1980). Genic and organismic selection. *Evolution* **34**, 825–843.

Wright, S. (1982). The shifting balance theory and macroevolution. *Ann. Rev. Genet.* **16**, 1–19.

Wu, L. and Antonovics, J. (1976). Experimental ecological genetics in *Plantago*. II. Lead tolerance in *Plantago lanceolata* and *Cynodon dactylon* from a roadside. *Ecology* **57**, 205–208.

Wu, L., Bradshaw, A. D. and Thurman, D. A. (1975). The potential for evolution of heavy metal tolerance in plants. III. The rapid evolution of copper tolerance in *Agrostis stolonifera. Heredity* **34**, 165–187.

Wulff, R. D. (1986a). Seed size variation in *Desmodium paniculatum*. II. Effects on seedling growth and physiological performance. *J. Ecol.* **74**, 99–114.

Wulff, R. D. (1986b). Seed size variation in *Desmodium paniculatum*. III. *J. Ecol.* **74**, 115–122.

10 Eco-geographical Differentiation in Some Aquatic Plants: the Lemnaceae

ELIAS LANDOLT

Geobotanisches Institut ETHZ, Stiftung Rübel, Zürich, Switzerland

I. Introduction

Aquatic plants are specialized for life in the water. Compared with a terrestrial environment, the aquatic environment is characterized by special features:

(1) greater uniformity of temperature regime: this results in a larger distribution area of most hydrophytes, including different climatic zones;
(2) permanently sufficient water supply: no xeromorphic structures are needed;
(3) possibility of absorbing nutrients directly from the water by all the submerged parts of the plant: the transport of nutrients by roots from the soil can be neglected. However, in this case, a high nutrient concentration in the water is needed, except in special cases of carnivory or symbiosis with nitrogen-fixing bacterias;
(4) supporting of the plant body by the water: no strengthening tissue, no lignification is needed.

Most aquatic vascular plants developed relatively early (late Mesozoic or early Tertiary) from terrestrial plants. The result is a great variety of small and isolated water plant families, isolated from each other.

Families which contain both terrestrial and true water plants are not frequent. Sculthorpe (1967) lists 33 families consisting only of water plants, 30 of which include fewer than 10 genera, 17 of which are monogeneric, and three which are monotypic. The taxonomic position of many of these families is not clear

Differentiation Patterns in Higher Plants
ISBN: 0-12-709425-3

owing to the strong adaptive changes of many features. The duckweeds (Lemnaceae) are an excellent example of such a specialized aquatic family.

II. Principal Features of the Duckweeds

The relationship of the Lemnaceae to other families remains unclear though many studies have been undertaken to investigate possible homologies. Some botanists place the family in the vicinity of the superficially similar genus *Pistia* (Araceae) (e.g. Hegelmaier, 1868; Arber, 1919; Daubs, 1965; Meusel, 1951; Kandeler, 1979), while others see more affinities to the Alismatales or the Najadales (e.g. Lawalrée, 1945; Deyl, 1955). The question is still open. Recent results do not show any definite connection (see Landolt, 1986, Ch. 7).

TABLE 1

Hierarchic structure of the family of *Lemnaceae* (number of species given in brackets)

Family	Lemnaceae (34)			
Subfamilies	Lemnoideae (16)		Wolffioideae (18)	
Genera	*Spirodela* (3)	*Lemna* (13)	*Wolffiella* (9)	*Wolffia* (9)

The family of Lemnaceae consists of two subfamilies, four genera, and 34 species (Table 1). Among the water plants, the Lemnaceae fill a narrow ecological niche: they live as very small free-floating pleustophytes, neither rooting in the soil nor rising above the water surface. They approximate phytoplankton organisms. In the extreme, the plant body is reduced to a small and undifferentiated green ovoid called a *frond* measuring about 0.5 mm in diameter with an epidermis, an assimilation tissue, and a meristematic point in the centre forming new fronds or flowers (Fig. 1). Generative propagation is rare in most species. The flowering organs of *Wolffia* are reduced to one ovary with a single ovule and one stamen with a two-loculated anther.

Typical characteristics of all Lemnaceae are:

- smallness;
- high mobility;
- small absorbing surface compared to biomass (no root hairs);
- large assimilating surface compared to biomass;
- simple anatomical structure;
- fast vegetative growth (doubling within 24 h at best).

Though the plant body is morphologically reduced in many respects, the physiological and phytochemical processes are essentially the same as in other vascular plants. For instance, the number of known enzymes in Lemnaceae exceeds 100 (Landolt and Kandeler, 1987).

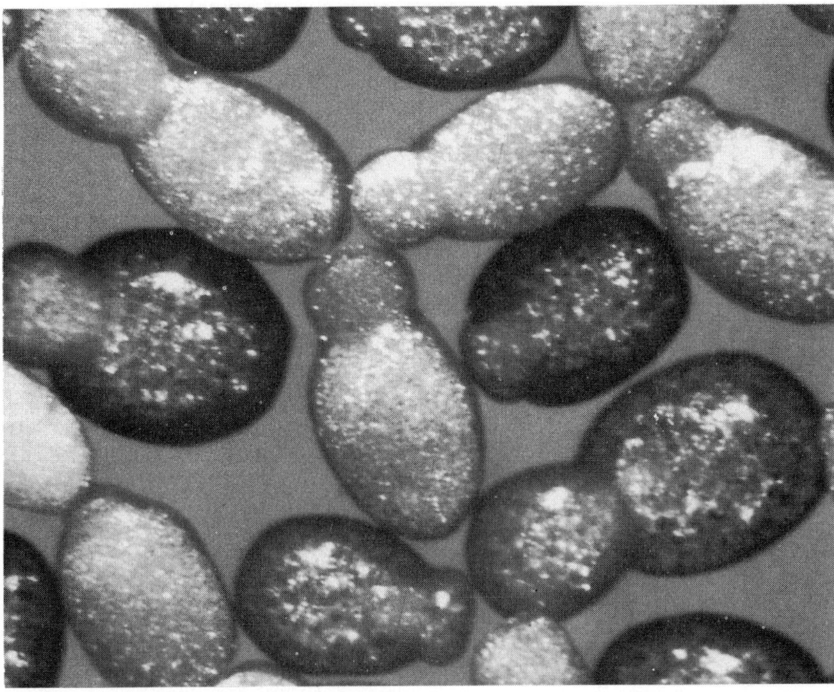

FIG. 1 Two species of *Wolffia*, the genus including the smallest flowering plants: *Wolffia borealis* (light ponds) and *Wolffia columbiana* (dark ponds); × 25. Photograph courtesy of W. P. Armstrong, San Marcos, CA, U.S.A.

The main advantages which lemnoid growth form provide are as follows:

- the possibility of filling up gaps within the vegetation cover of the water;
- the capability for rapid distribution within a water body and from water body to water body.

The disadvantages are:

- the need for high nutrient concentration in the water;
- the lack of tolerance to water movements (wind, waves, flow);
- the lack of the ability to compete for light with emergent plants.

III. Speciation within the Family of Lemnaceae

Since the Lemnaceae mostly propagate vegetatively, speciation through mutation plays a more important role than through recombination. Even if a given species produces flowers frequently, the single flower is surrounded by flowering fronds of the same vegetatively spread clone, and seeds are formed mostly after self-pollination. On the other hand, the rare occurrence of sexual propagation may effectively isolate an occasional hybrid. Two species of possible hybrid origin are known within the family of Lemnaceae: *Lemna japonica* and *L. perpusilla*. Both are morphologically intermediate and occur eco-geographically between the supposed parents (Fig. 2).

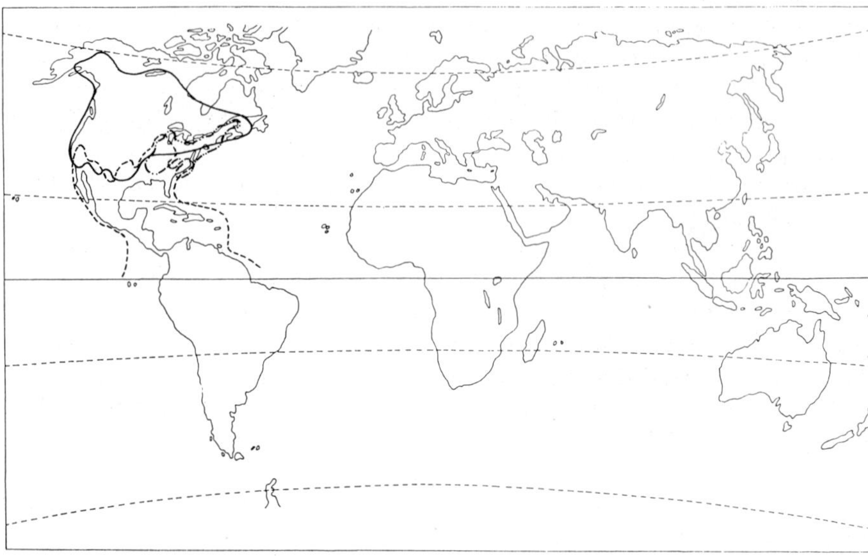

FIG. 2 Distribution areas of *Lemna perpusilla* (-···-) and its supposed parent species *L. turionifera* (——) and *L. aequinoctialis* (---) in North America.

The fast vegetative propagation (doubling within 24 h at best) enables a population to form many mutants within a relatively short time. If we arbitrarily assume a total of 10 000 genes in each chromosome set and an average mutation rate of 10^{-5} mutations per generation and gene, mutations should occur within 10% of the fronds per generation. Assuming a vegetation period throughout the year and the maximum growth rate, the number of individuals starting from only one reaches 2^{250} (about 10^{45}) after a year. This means that theoretically 10^{44} mutations occurred during the year and fronds with up to at least nine mutated genes could originate. It is therefore not astonishing that within the Lemnaceae effective and fast differentiation is possible even if no recombinations through crossing occur. The variation of chromosome numbers within the family is remarkable. Within most species Urbanska (1980) counted different ploidy levels as well as aneuploid numbers. She found no distinct correlation between chromosome number and climatic or geographic region. Geber (pers. comm.), applying DNA specific fluorochromes to mark the heterochromatin, observed some differentiation within the genus *Spirodela* ($2n = 36$ for *S. intermedia*,

$2n = 40$ for *S. polyrrhiza* and $2n = 46$ for *S. punctata*), whereas the other genera did not show much chromosomal differentiation (predominantly $2n = 42$).

Though the Lemnacean plant body is very small and distinctive characteristics rather few, genetically based variation in the Lemnaceae has been known for many years (e.g. Mendiola, 1919; Landolt, 1957; Spooner, 1967; Hillman, 1979; Porath *et al.*, 1980; Beppu and Takimoto, 1981b,c; Davidson and Simon, 1981; Landolt and Dann, 1983; for a complete list of publications see Landolt, 1986). Besides morphological features, these differences comprise growth rates, responses to climatic and nutrient conditions, flowering behaviour, enzyme activities, etc. In some investigations, a correlation could be detected between the clonal behaviour and environmental factors at the place of origin (e.g. temperature, content of heavy metals in the water, X-radiation).

An interesting example of intraspecific differentiation was detected by Beppu and Takimoto (1981a,b, 1983) and Beppu *et al.* (1985). *L. aequinoctialis*, a species of tropical and subtropical areas of the world, can be divided in Japan into four different taxa according to growth behaviour and some minor morphological features (Table 2). In central and northern Japan, *L. aequinoctialis* is almost entirely restricted to rice fields. Compared to other continents, (e.g. western Asia, South America, Australia) it became spread with the cultivation of rice far beyond the natural distribution area towards the polar circle. This specialization as a weed of rice culture and the successive adaptation with the extension of these cultures to northern areas to survive cooler winters is an example of a rather recent racial differentiation. Outside Japan, *L. aequinoctialis* shows a certain differentiation pattern (Beppu, pers. comm.). From 21 investigated strains from all over the world, Beppu found seven different DNA patterns, partly connected with geographical areas. A more extensive study might reveal some interesting aspects on the world-wide differentiation of the species.

TABLE 2

Differentiation of *L. aequinoctialis* in Japan, according to Beppu *et al.* (1985)

	Taxa			
Characteristics	N_1	N_2	K	S
Hibernating organ	Seed	Turion	Frond	Frond and seed
Flowering behaviour	Short day	Short day	Day neutral	Long day
Growth rate	Low	Low	Low	Low
Chromosome number ($2n$)	*c.* 70	40	50	40
Distribution in Japan	Centre and north	West coast of central Honshu	Kyoto (introduced)	South

IV. Examples of Eco-geographical Differentiation within Genera of the Lemnaceae

South America is the centre of diversity of the Lemnaceae and must be regarded as the place of origin. The species with most original characteristics of all four genera of Lemnaceae have their main distribution in South America (Table 3).

Today, the most differentiated plants within the family of Lemnaceae belong to *Spirodela intermedia*. In all essential characteristics this species seems to be the most primitive; therefore it is thought to be near the origin of Lemnaceae development. It has the biggest fronds (up to 0.8 cm^2), highest number of roots and nerves, largest tracheid system, large aerenchymatic tissue, several ovules per ovary, diverse flavonoid pattern, etc. The species is restricted to South America (and a few places in Central America), and grows in regions with tropical to warm temperate climates where frosts are at most slight and rare.

The ecological differentiation followed mainly four different lines.

(1) Reduction in size and subsequent simplification of the whole structure: the fronds become more mobile and better adapted to fill gaps within the vegetation cover. Two different growth forms seem to be especially successful: (a) small flat to gibbous ovate fronds with a stabilization organ (root, appendage) and of a medium size of about 3 mm to 2 mm (one species of *Spirodela*, nine species of *Lemna*, one species of *Wolffiella*); (b) very small ovoid bodies of 0.5 – 1.5 mm in diameter (seven species of *Wolffia*). Both types of fronds are distributed world-wide.

(2) Adaptations to growing at lower temperatures and to tolerate frosts. This tendency is restricted to *Spirodela punctata* and the genus *Lemna* (all species except *L. aequinoctialis* and possibly *L. ecuadoriensis* as well as *L. tenera*).

(3) Formation of special resting buds (turions) which sink to the bottom of the water. In this way, the plant is able to survive unfavourable conditions of temperature and nutrient content. Turions are achieved in *Spirodela polyrrhiza*, in *Lemna turionifera* (and partly in *L. aequinoctialis*) and in the genus *Wolffia* (all species except *Wolffia elongata* and *Wolffia microscopica*). In *S. polyrrhiza* and in *Lemna* the turions are moderately frost-resistant; in *Wolffia* they have the same sensitivity to frost as the normal fronds.

(4) Development of a submerged life form. The temperature conditions below the water surface are less extreme, and the nutrient content more favourable due to a nutrient gradient from soil to the water surface. Submerged life is achieved independently in *L. trisulca*, *L. tenera*, and partly in *L. valdiviana*, as well as in most species of *Wolffiella*. Plants growing below the water surface are easily overgrown by species floating on the surface and get less light. To coexist with such plants they are highly shade tolerant and heterotrophically able to use organic substances dissolved in the water. If a nutrient in the water is in short supply, they have the best chance in competition since they are first able to take up the newly released nutrients from the soil.

TABLE 3
Number of species in different continental areas

Continential area	*Spirodela*			*Lemna*			*Wolffiella*			*Wolffia*			*Lemnaceae*		
	1	2	3	1	2	3	1	2	3	1	2	3	1	2	3
South America	3	1	—	5	1	—	4	1	—	5	1	—	17	4	—
North America	2	—	1	9	2	—	3	1	—	4	1	—	18	4	2
Africa	2	—	—	4	—	—	5	4	—	2	—	—	13	4	—
Europe and West Asia	2	—	1	5	—	1	—	—	—	1	—	—	7	—	2
East and South Asia	2	—	—	8	2	2	1	—	1	4	1	—	15	3	3
Australia	2	—	—	4	1	1	—	—	—	2	1	—	8	2	1
World-wide	3	1		13	6		9	6		9	4		34	17	

1, total number of species; 2, number of endemic species; 3, number of introduced species.

Some examples will illustrate this ecological differentiation.

A. Spirodela polyrrhiza

S. polyrrhiza is the nearest relative of *S. intermedia*. In South America it is restricted to the northwest. The main difference from *S. intermedia* is the ability to form turions. The turions enable the species to spread to regions that are cold in winter. In South America, the competition of the very similar *S. intermedia* prevented the spreading to the south. Since the lowest growing temperature for *S. polyrrhiza* is about 15°C (Landolt, 1957), the species avoids regions with cool summers (summer temperatures below 16°C). Very dry regions (factor of Martonne smaller than three) are not colonized by *S. polyrrhiza*. Spreading direction was from South America to Central and North America and from there to Europe, western Asia and Africa as well as to eastern Asia and Australia (Fig. 3).

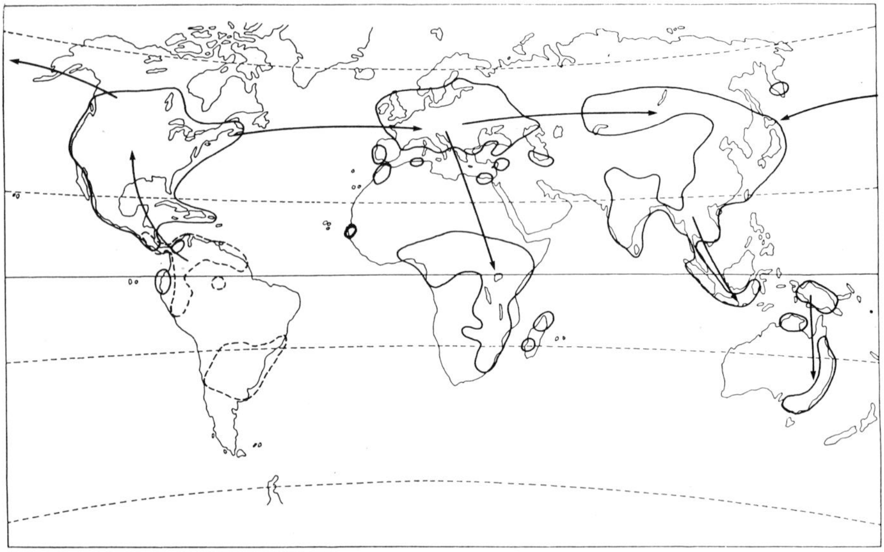

FIG. 3 Distribution areas of *Spirodela intermedia* (---) and *S. polyrrhiza* (——) with supposed directions of expansion.

B. Lemna gibba

L. gibba is the most primitive of the genus *Lemna*. It has still more than one ovule per ovary and the biggest frond areas of the *Lemna* species. Although different from *S. intermedia*, the species is still able to grow at temperatures of down to about 4°C, and the fronds are moderately frost- but less heat-tolerant (Landolt, 1957). They are more tolerant to high salt concentrations and can survive dry periods as seeds. On the other hand, they need a higher nutrient concentration than most other species (e.g. Lüönd, 1983; Landolt and Kandeler, 1987, Ch. 2).

These new characteristics allowed the species to spread to drier and moderately winter-cold areas, but they cannot be found in hot and relatively humid regions (factor of Martonne higher than four). They spread from South America north to California and west to South Africa, and from there to East Africa and Europe and to Australia. The isolated populations of southern Australia (and New Zealand) show some statistical differences from the other populations and are called *L. disperma* (Fig. 4).

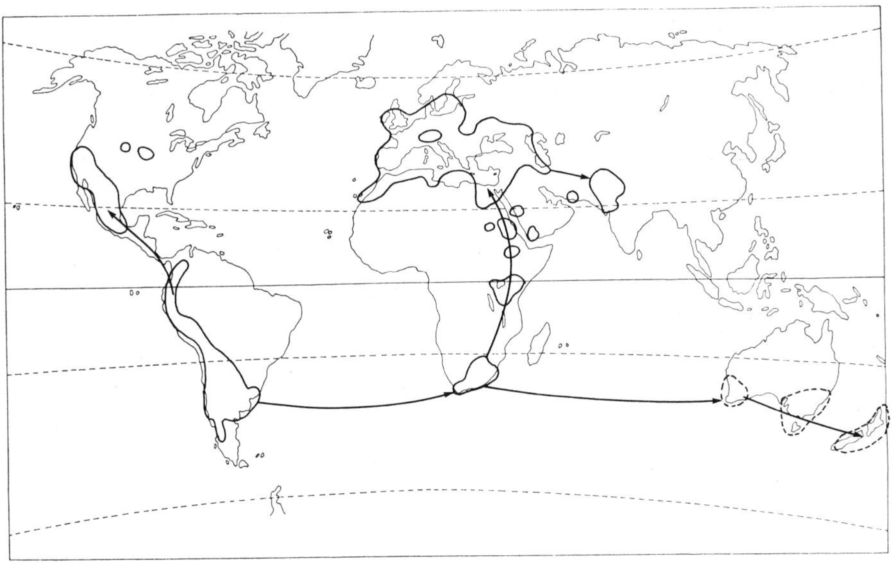

FIG. 4 Distribution areas of *Lemna gibba* (——) and *L. disperma* (---) with supposed directions of expansion.

C. Lemna turionifera – L. minor – L. obscura

Since the relatively high need of nutrients of *L. gibba* and the inability to tolerate hard frosts prevented *L. gibba* from colonizing both the more continental as well as the more humid areas of North America, these climatic conditions caused an ecological differentiation. The formation of turions enabled plants to spread to continental areas with cold winters (*L. turionifera*) in North America and eastern and central Asia. Within the more humid eastern part of North America two new species differentiated, one (*L. obscura*) in warm temperate regions and one (*L. minor*) in cool regions. *L. minor* was able to spread to Europe and from there to western Asia and Africa. *L. obscura*, however, could not reach the nearest region with similar climate in eastern Asia and remained endemic. Instead, a new species (*L. japonica*) developed in that region. This species probably originated from a hybrid between *L. turionifera* and *L. minor*, which either are not competitive with *L. japonica* today or have not yet been detected in the area due to the relatively difficult distinction between them (Fig. 5).

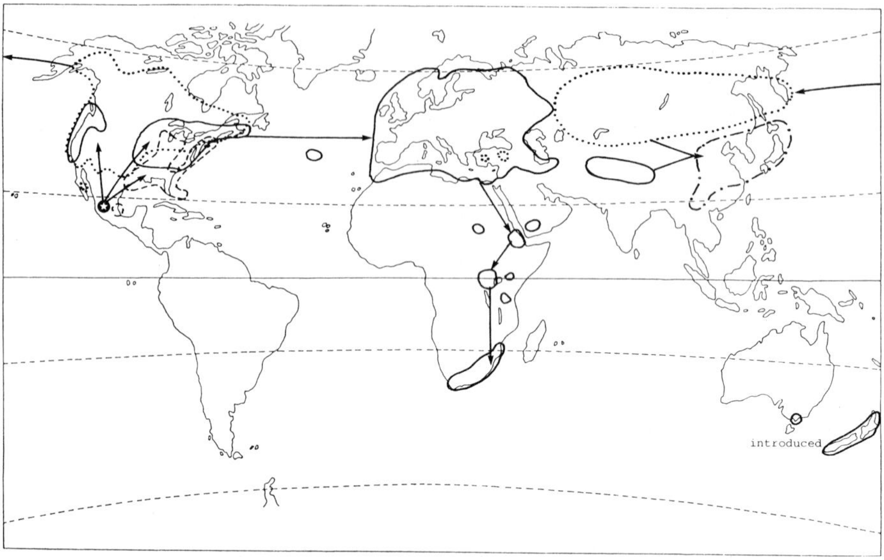

FIG. 5 Distribution areas of species of the *Lemna minor* group with supposed directions of expansion; (·····) *L. turionifera*; (---) *L. obscura*; (——) *L. minor*; (-·-) *L. japonica*. *Area of supposed centre of differentiation.

D. Lemna trisulca

L. trisulca derived from an ancestor within the *L. gibba–L. minor* group. The submerged life form gave the species the ability to spread to cool areas. A simultaneous loss of tolerance to high temperature prevented the species from colonizing regions with warm summers. From North America, the species spread to East Asia – Japan – the Philippines – New Guinea and Australia on one side, and to Europe–Western Asia and East Africa on the other side. A slight morphological differentiation took place in the isolated populations of Australia, where roots are formed very rarely, and one nerve instead of three is normally present. However, since the presence of these characteristics is not restricted to Australia and roots as well as three nerves have been observed in some herbarium specimens from Australia, no separation of a distinct taxon is given (Fig. 6).

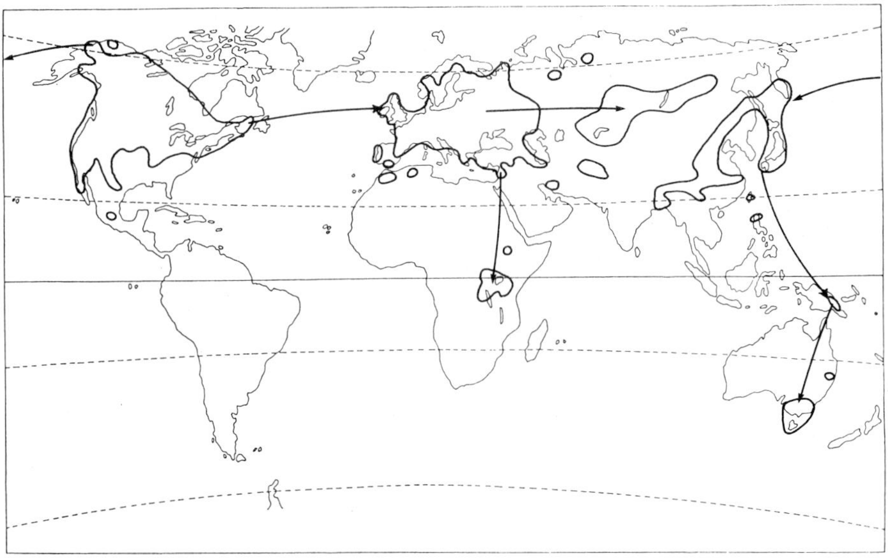

FIG. 6 Distribution area of *Lemna trisulca* (——) with supposed directions of expansion.

E. Wolffiella

Within the genus *Wolffiella* no frost resistance developed. *W. neotropica* and *W. welwitschii* are typical species of the genus (Fig. 7). The few species which are able to survive in regions with frosts live below the surface of the water. The ability to grow at temperatures down to 10°C is the reason for the spreading of *W. gladiata, W. oblonga* and *W. denticulata* to regions with temporary cool periods. *W. neotropica*, which is considered to be the most primitive species of *Wolffiella*, grows sometimes on and sometimes below the water surface. An African branch of the genus which probably originated from *W. neotropica* always lives on the surface of the water and does not grow outside the tropical and subtropical regions. It developed only in Africa and formed three endemic species (Fig. 8).

F. Wolffia

With two exceptions (*Wolffia microscopica, Wolffia elongata*), all species of *Wolffia* are able to produce turions. Therefore, even if they do not grow at temperatures much below 15°C, they can survive in regions with cool winters. The most original species, *Wolffia brasiliensis*, has an American distribution. One branch of differentiation leads to the group of *Wolffia arrhiza. Wolffia arrhiza* s.str., the original species of the group, grows in the region of Rio de Janeiro and in Africa, Europe, and western Asia.

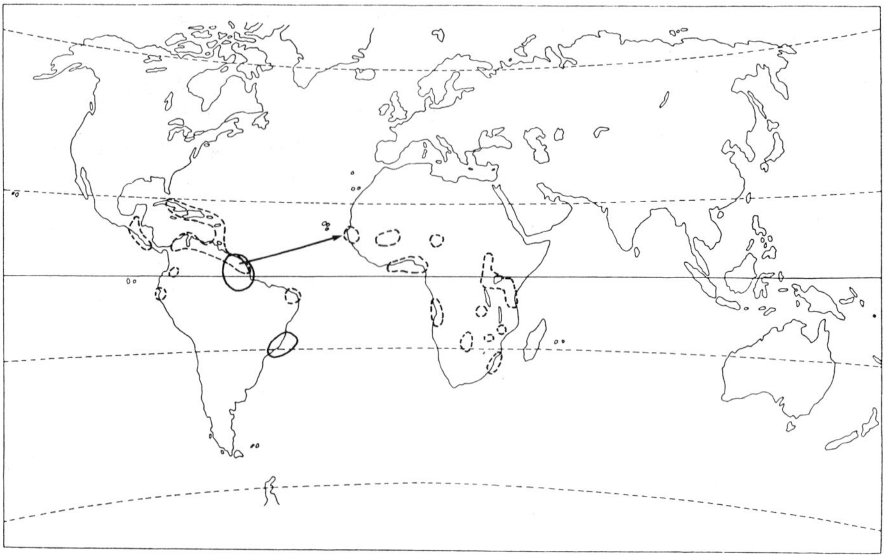

FIG. 7 Distribution of *Wolffiella neotropica* (——) and *W. welwitschii* (---) with supposed directions of expansion.

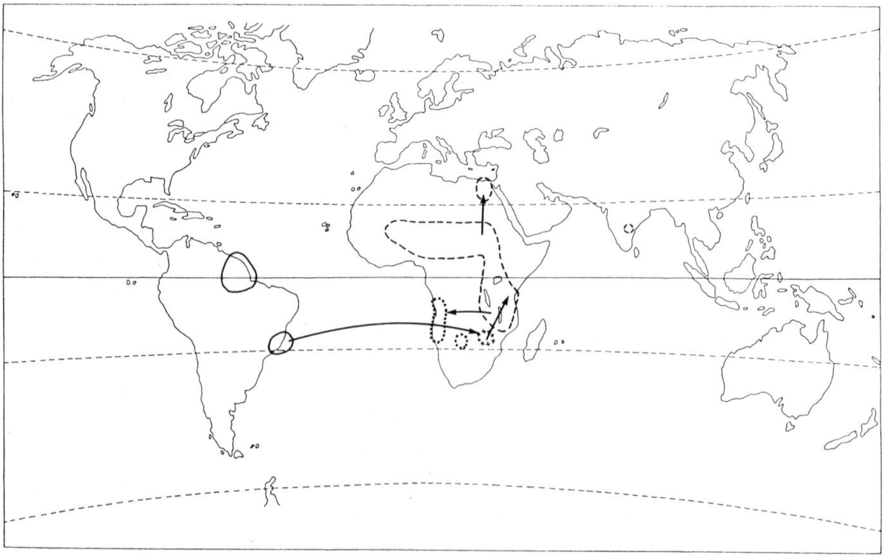

FIG. 8 Distribution areas of *Wolffiella neotropica* (——), *W. rotunda* (+++), *W. hyalina* (---) and *W. repanda* (·····) with supposed directions of expansion.

Wolffia arrhiza and *Wolffia brasiliensis* are distinguished by few morphological characteristics, but having a very similar life form they exclude each other. *Wolffia arrhiza* gave rise to a mainly geographical differentiation: two species in Australia, one species in East Asia, and one species in America, with similar nutrient and climatic demands. Due to a slightly submerged life form, *Wolffia columbiana* is able to coexist in America with *Wolffia brasiliensis* and has almost the same distribution (Fig. 9).

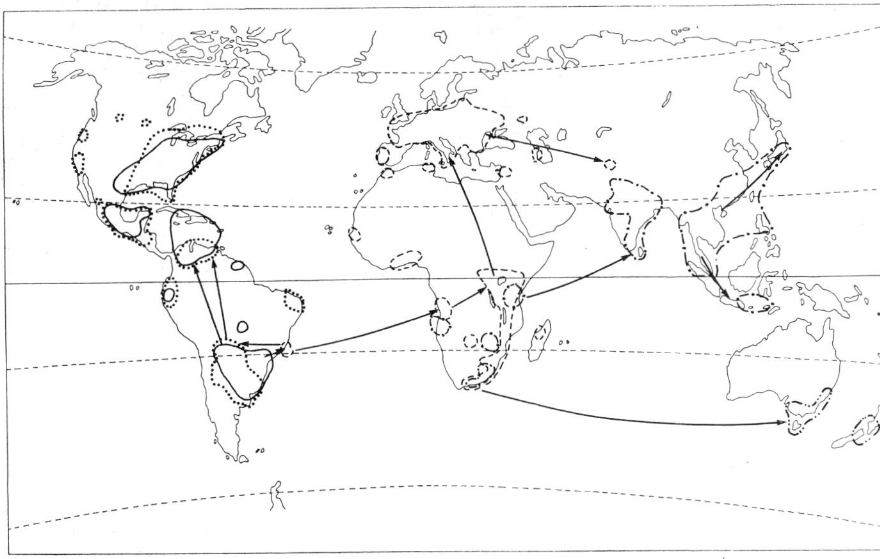

FIG. 9 Distribution areas of members of the genus *Wolffia* with supposed directions of expansion (——) *W. brasiliensis*; (---) *W. arrhiza*; (-·-) *W. globosa*; (-···-) *W. australiana*; (·····) *W. columbiana*.

V. Conclusions

The *Lemnaceae* are a family of genuine water plants that evolved in quite a special way, the most prominent characteristics being reduction in size and morphology. The duckweeds are the smallest flowering plants floating in water without any direct contact with the soil; their way of life approximates that of phytoplankton organisms. This extreme specialization makes it probable that the distribution pattern of the Lemnaceae is not typical for all water plants. On the other hand, the eco-geographical differentiation is quite distinct, contrary to the general opinion that duckweeds are ecologically and geographically ubiquitous. Certainly, the Lemnaceae are easily distributed by birds over short distances and life conditions in water are not as different world-wide as are those in terrestrial systems. Nevertheless, temperature conditions and chemical composition of the water all over the world vary enough to permit a characteristic differentiation.

Patterns of geographical distribution of the *Lemnaceae* strongly suggest that the centre of origin of the family is situated in the warm temperate to subtropical zones of South America. *Spirodela intermedia*, which is the most differentiated and therefore the most primitive species of the *Lemnaceae*, fits into this climate

Ecological differentiation within the *Lemnaceae* comprises both life-history elements as well as development of particular tolerances. Four principal pathways have been recognized.

(1) Production of specialized resting buds, namely turions, which sink to the bottom of the water and survive there in unfavourable conditions (e.g. lack of nutrients, cold winters). In a few taxa, seeds with the same behavioural pattern are produced.
(2) Physiological ability to grow at cool temperatures and development of frost resistance.
(3) Production of drought-resistant seeds enabling a given taxon to colonize seasonal waters.
(4) Development of submerged habit in order to avoid extreme temperatures and to ensure a better access to the nutrients newly released from the sediments.

All four pathways have appeared several times during the development of the duckweed family. In this way the *Lemnaceae* have succeeded in colonizing nearly all climates and regions of the world except the areas with cold summer climate (average temperatures below 10°C) and with very dry and exceedingly wet climates (too high or too low concentrations of some chemical compounds in the water).

References

Arber, A. (1919). On the vegetative morphology of *Pistia* and the Lemnaceae. *Proc. Roy. Soc.* **91**, 96–103.

Beppu, T. and Takimoto, A. (1981a). Geographical and cytological variation of *Lemna paucicostata* Hegelm. in Japan. *Bot. Mag. Tokyo* **94**, 11–20.

Beppu, T. and Takimoto, A. (1981b). Further studies on flowering of *Lemna paucicostata* in Japan. *Bot. Mag. Tokyo* **94**, 69–76.

Beppu, T. and Takimoto, A. (1981c). Growth of various ecotypes of *Lemna paucicostata* in Japan under various temperature conditions, and their wintering forms. *Bot. Mag. Tokyo* **94**, 101–114.

Beppu, T. and Takimoto, A. (1983). Flowering behavior of the hybrids between strains 6746 and 371 of *Lemna paucicosfata* Hegelm. *Aquat. Bot.* **17**, 295–300.

Beppu, T., Yanase, D., Nobuchi, T. and Murata, G. (1985). A revision of *Lemna paucicostata* group in Japan. *Acta Phytotax. Geobot.* **36**, 45–58 (in Japanese, English summary).

Daubs, E. H. (1965). A monograph of *Lemnaceae. Illinois Biol. Monogr.* **34**.

Davidson, D. and Simon, J. P. (1981). Thermal adaptation and acclimation of ecotypic populations of *Spirodela polyrrhiza* (L.) Schleid. (Lemnaceae); thermostability and apparent activation energy of NAD malate dehydrogenase. *Can. J. Bot.* **59**, 1061–1068.

Deyl, M. (1955). The evolution of the plants and the taxonomy of the monocotyledons. *Acta Mus. Nat. Pragae* **11B**(6).

Hegelmaier, F. (1868). "Die Lemnaceen. Eine monographische Untersuchung". Engelmann, Leipzig.

Hillman, W. S (1979). Temporal compartmentation in *Lemna paucicostata* Hegelm.: photoperiodism, respiration, nitrogen nutrition and beterotrophic growth of different strains. *Am. J. Bot.* **66**, 1021–1028.

Kandeler, R. (1979). Familie: Lemnaceae. *In* "Illustrierte Flora von Mitteleuropa", Vol. II/1,5 (G. Hegi, ed.), pp. 335–346. Parey, Berlin/Hamburg.

Landolt, E. (1957). Physiologische und ökolgische Untersuchungen an Lemnaceaen. *Ber. Schweiz. Bot. Ges.* **67**, 271–410.

Landolt. E. (1986). The family of *Lemnaceae* — a monographic study. 1. *Veröff. Geobot. Inst. ETH, Stiftung Rübel, Zürich* **71**, 566pp.

Landolt, E. and Dann, W. (1983). Vergleich von 10 Klonen von *Lemna gibba* bei verschiedenen Stickstoffkonzentrationen. *Ber. Geobot. Inst. ETH, Stiftung Rübel, Zürich* **50**, 86–97.

Landolt, E. and Kandeler, R. (1987). The family of *Lemnaceae* — a monographic study. 2. *Veröff. Geobot. Inst. ETH, Stiftung Rübel, Zürich* **95**.

Lawalrée, A. (1945). La position systématique des Lemnacées et leur classification. *Bull. Soc. R. Bot. Belg.* **77**, 27–38.

Lüönd, A. (1983). Das Wachstum von Wasserlinsen (Lemnaceae) in Abhängigkeit des Nährstoffangebots, insbesondere Phosphor und Stickstoff. *Veröff. Geobot. Inst. ETH, Stiftung Rübel, Zürich* **80**, 116.

Mendiola, N. B. (1919). Variation and selection within clonal lines of *Lemna minor. Genetics* **4**, 151–182.

Meusel, H. (1951). Die Bedeutung der Wuchsform für die Entwicklung des natürlichen Systems der Pflanzen. *Feddes Rep.* **54**, 137–172.

Porath, D. Efrat, I. and Arzee, T. (1980). Morphological patterns and heterogeneity in populations of duckweeds. *Aquat. Bot.* **9**, 159–168.

Sculthorpe, C. D. (1967). "The Biology of Aquatic Vascular Plants". Edward Arnold, London.

Spooner, W. E. (1967). Local Population variation in *Spirodela polyrrhiza* (L.) Schleid. (Lemnaceae). MS Thesis, North Carolina State University, Raleigh, NC.

Urbanska, K. (1980). Cytological variation within the family of Lemnaceae. *Veröff. Geobot. Inst. ETH, Stiftung Rübel, Zürich* **70**, 30–101.

11 Vegetative Growth and Genetic Mobility in Some Aquatic Weeds

CHRISTOPHER D. K. COOK

Institut für Systematische Botanik der Universität Zürich, Zürich, Switzerland

I. Introduction

In the last few years some interesting correlations between breeding systems and ecological or eco-geographical factors have been revealed (see Jain, 1976; Willson, 1979; Givnish, 1980; Bawa and Beach, 1981; Sobrevila and Kalin Arroyo, 1982; Baker and Cox, 1984). I thought it worthwhile to see if aquatic plants as a group are in any way different (in terms of breeding system) from other groups of plants such as tropical forest trees, those growing on oceanic islands, prairie, etc. Also, it was felt worthwhile to combine data on breeding systems with dispersal mechanisms.

II. The Environment

It is often stated that the environment in which aquatic plants grow is very uniform compared to the terrestrial environment (Sculthorpe, 1967; see also Landolt's contribution in this volume). This uniformity is often considered to be the reason why so many aquatic plants are widely distributed. Although it is true that many aquatic species do occupy very large areas, it is equally true that many do not, as shown by Cook (1983) in a review of the aquatic plants endemic to Europe and the Mediterranean. Even in large and apparently uniform areas such as the rice fields along the coastal plain of Kerala, India, several localized endemic genera and species are found that occupy well-defined ranges within the region. Cook (1980) discussed the distribution of the following species endemic to southwest

Differentiation Patterns in Higher Plants
ISBN: 0-12-709425-3

India: *Aponogeton appendiculatus* van Bruggen, *Limnopoa meeboldii* (C. E. C. Fischer) C. E. Hubbard, *Nymphoides macrospermum* R. Vasudevan Nair, *Rotala macrandra* Koehne, *Rotala malampuzhensis* R. Vasudevan Nair, and *Wiesneria triandra* (Dalzell) Micheli.

Within single bodies of water one tends to find (exactly as in terrestrial communities) that the individual species are distributed in complex patchworks in which individuals or clones of individuals exist side by side or are intermingled. To date no reliable and reproducible results have been obtained from experimental work on how aquatic plant species or clones influence their neighbours. Nevertheless, many authorities are convinced that allelopathic mechanisms exist. Uniformity of the environment may explain the large-scale distribution of many aquatics, when measured in kilometres, but on a small scale, measured in metres or millimetres, the aquatic environment, like the terrestrial one, is a mosaic.

On a world-wide scale it is estimated that there are about 5000 species of aquatic macrophytes (Cook *et al.*, 1974). Our knowledge of the breeding system, dispersal mechanism and ecology of most of these species is so poorly known that it is not yet possible to consider all aquatics throughout the world.

Man, through travel and his talent for disturbing the environment has, among other things, provided aquatic plants not only with the means of transport, but has also supplied suitable areas for colonization (artificial lakes, canals, ditches, etc.). Many aquatics have recently extended their range. In a review, Cook (1985) considers the spread of 198 aquatic species of which 172 are reasonably well documented. From a biosystematic point of view, the best known aquatics are those that interfere with man and have earned the title "weed". Weeds are usually worse pests outside their native range. However, they offer the evolutionist the opportunity to study the behaviour of species in rather different environments. I have chosen 12 of the most notorious aquatic weeds to compare breeding system and genetic mobility.

III. The Plants

A. Trapa natans *L. sensu lato (Trapaceae), Water chestnut*

Bottom-rooted with leaves in floating rosettes; obligate annual (monocarpic). Native in the Old World, introduced into North America; cultivated for food (edible fruits) in parts of Asia and Europe, pest in North America and parts of Asia (Caspian Sea). Flowers bisexual, pollination by insects or more commonly autogamous; usually self-fertilized but some chance for cross-fertilization, although insects rarely visit the flowers. Effective spread by seed only.

B. Najas minor *All. (Najadaceae), Naias*

Bottom-rooted, totally submerged with underwater pollination; annual or perennial (often phenotypically annual in temporary water). Native in Eurasia and north Africa, introduced in North America; pest in North America and parts of Eurasia, particularly in rice fields. Flowers unisexual; plants monoecious; probably mostly self-fertilized but good chances for cross-fertilization; seed-set usually very good. Effective spread by seeds and vegetative fragments.

C. Myriophyllum spicatum *L. (Haloragaceae), Eurasian milfoil*

Submerged, bottom-rooted and stoloniferous perennial with emergent flowers, it can grow terrestrially in summer. Native to northern Eurasia; introduced into North America where it is often considered to be a pest, see Anderson (1985). Flowers unisexual (rarely a few bisexual); plants monoecious. Wind-pollinated, self-compatible, mostly self-fertilized but good chances for cross-fertilization. Effective spread by vegetative fragments and seed.

D. Pistia stratiotes *L. (Araceae), Water lettuce*

Free-floating, annual or perennial, rosettes with well-developed stolons. Pantropical, it is not possible to distinguish native and introduced ranges. Flowers unisexual; plants moniecious; insect or self-pollinated; mostly self-fertilized but some chances for cross-fertilization. Effective spread by seed or vegetative fragments.

E. Eichhornia crassipes *(Mart.) Solms. (Pontederiaceae), Water hyacinth*

Free-floating, annual or perennial rosettes with well-developed stolons. Native in the central part of tropical South America, introduced in warmer parts of all continents, often regarded as the most troublesome weed in the world (Gopal and Sharma, 1981). Flowers bisexual, insect pollinated, tristylic, weakly self-compatible; middle style morph most widely distributed, the pin style morph less so and the thrum confined to America. Effective spread by vegetative fragments and seed.

F. Alternanthera philoxeroides *(Mart.) Griseb. (Amaranthaceae), Alligator weed*

Emergent perennial, rooting in the substrate or floating on the surface. Native in southern tropical and subtropical South America, introduced in Central and North America, Asia and Australia, where it is sometimes a serious pest in canals and drainage ditches. Flowers bisexual, probably insect pollinated, probably self-incompatible because seeds are unknown outside its native range. Effective spread by vegetative fragments.

G. Salvinia molesta *Mitchell (Salviniaceae), Giant salvinia*

Free-floating, perennial with regularly branching stems. Probably a hybrid originating in tropical South America, introduced in Africa, Asia and Australasia, a serious pest in many regions. Heterosporic fern with disturbed meiosis, sexual reproduction never observed. Effective spread by vegetative fragments.

H. Egeria densa *Planchon (Hydrocharitaceae), Egeria or Brazilian elodea*

Rooted submerged perennial with emergent flowers. Native in southeastern South America, introduced in western and northern South America, Central and North America, Europe, eastern and southern Asia, a serious pest in North America and New Zealand. Flowers unisexual, insect pollinated; plants dioecious, female flowers unknown until 1981, see Cook and Urmi-König (1984), seeds unknown in nature but have developed in University of Zürich Botanic Garden, see Cook and Urmi-König (1984). Effective spread by vegetative fragments of male plants.

I. Lagarosiphon major *(Ridley) Moss (Hydrocharitaceae), Lagarosiphon or African elodea*

Rooted submerged perennial with emergent flowers. Native in southern Africa

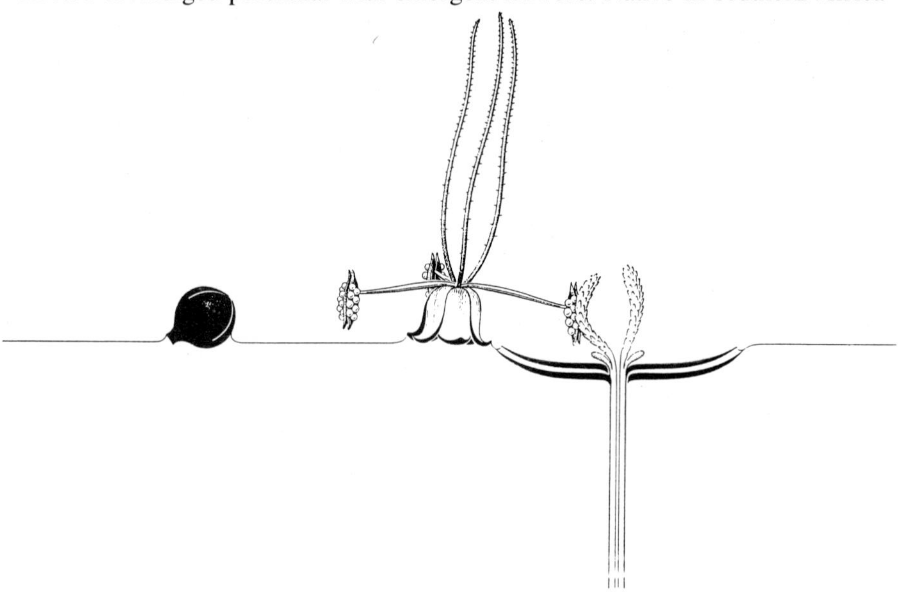

FIG. 1 Diagrammatic representation of the pollination of *Lagarosiphon*. Left, male flower bud; centre, male flower with erect staminodia; right, female flower.

(Zimbabwe and RSA), introduced in Europe and New Zealand, serious pest in waterways of New Zealand. Flowers unisexual, pollination on the water surface with free-floating male flowers (Fig. 1); for details see Cook (1982). Plants dioecious, female plants only known outside native range. Effective spread by vegetative fragments of female plants.

J. Elodea canadensis Michaux (Hydrocharitaceae), Canadian pondweed

Rooted submerged perennial with emergent flowers. Native in temperate North America, introduced in Europe, Australia and New Zealand; for details see Cook and Urmi-König (1985). In Europe it is no longer considered a pest but is a serious pest today in Australia. Flowers unisexual, pollinated at the water surface with free-floating pollen (Cook and Urmi-König, 1985); plants dioecious, in Europe female plants only today but males persisted for 20–30 years without developing breeding populations; in Australia males only; in New Zealand originally females but apparently being replaced by males today but without developing breeding populations. Effective reproduction by vegetative fragments of both sexes, seeds unknown outside native range.

K. Elodea nuttallii *(Planchon) St John (Hydrocharitaceae), Nuttall's pondweed*

Rooted submerged perennial with emergent flowers. Native in temperate North America, introduced in Japan and Europe where it is apparently replacing *Elodea canadensis*. Flowers unisexual, pollinated at the water surface by free-floating male flowers (Fig. 2); for details see Cook and Urmi-König (1985). Plants dioecious, in Japan males only, in Europe males and females present but mostly allopatric, breeding populations unknown outside its native range. Effective spread by vegetative fragments of male and female plants.

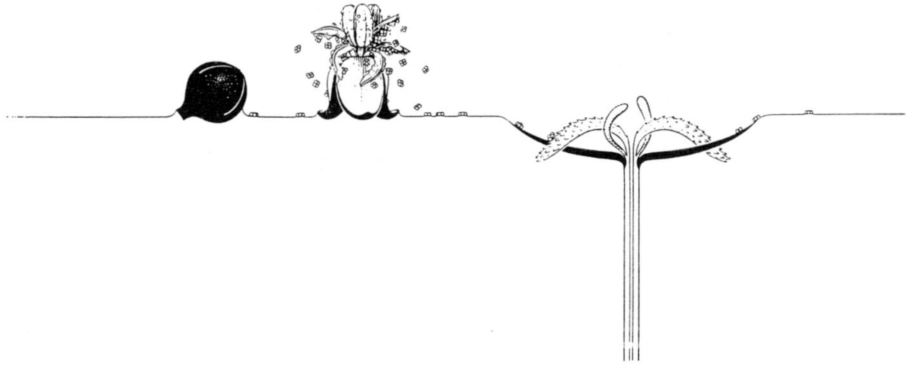

FIG. 2 Diagrammatic representation of the pollination of *Elodea*. Left, male flower bud; centre, male flower shedding pollen; right, female flower.

L. Hydrilla verticillata *(L. fil.) Royle (Hydrocharitaceae), Hydrilla*

Rooted submerged perennial with emergent flowers. Widely distributed in Old World but uncertain where it is truly native; probably the centre of origin lies in warmer regions of Asia; it has become established a long time ago (at least 200 years) in Europe and central Africa; in other parts of Africa it is a recent introduction and has become introduced in historical times to North, Central and northern South America, Pacific and Atlantic Islands, New Zealand and parts of Australia; it is a serious pest in southern U.S.A. Flowers unisexual, pollinated at water surface (Fig. 3); see Cook and Lüönd (1982). Plants female or monoecious (no reliable records of exclusively male plants); in Europe and the southern part of the U.S.A., northeastern India and Japan sterile triploid plants are recorded. Effective spread by vegetative fragments and specialized buds (tubers, turions or hibernacula), rarely by seed.

FIG. 3 Diagrammatic representation of the pollination of *Hydrilla*. Left, male flower bud; centre left, male flower "exploding"; centre right, female flower; right, "spent" male flower.

IV. Discussion

Trapa natans is the only species on this list that is monocarpic and without any means of vegetative spread. It is largely, when not exclusively, autogamous but, nevertheless, it is the only species dependent upon a sexual process for its reproduction and spread. All the other species have or can spread by means of vegetative fragments. Of these most have no specialized adaptations such as dormant buds, turions or tubers (they are mostly tropical or subtropical plants that do not have to overwinter), but in some cases such as *Hydrilla verticillata* specialized buds with distinct abscission layers develop.

The only other species that regularly develop seeds in their native and adventive ranges are *Najas minor, Myriophyllum spicatum* and *Pistia stratiotes*. However, all these species are capable of vegetative spread and seed is probably only important in plants growing in temporary water. These species are all autogamous and show no specialized outbreeding mechanisms.

At the other extreme is *Salvinia molesta*, which is a pentaploid with a disturbed meiosis (Loyal and Grewal, 1966; Schneller, 1981) and probably incapable of repeated sexual reproduction. Although it has no specialized buds or bulbils and shows very little resistance to cold and desiccation, it has spread and become a serious pest in Africa, southern and southeastern Asia and Australia. As the entire species is apparently a single clone whose population dynamics depend entirely on branching growth and fragmentation, it seems that it can be successfully controlled by a single vector, a weevil, *Cyrtobagous salviniae* Calder and Sands (Room, 1986).

Alternanthera philoxeroides develops no seed in its adventive range. It is presumably represented by a single genotype and is successfully controlled by a Flea-beetle, *Agasicles hygrophila*, and a pyralid moth, *Vogtia malloi*. *Eichhornia crassipes* is usually represented by a single genotype outside its native range and is successfully attacked by two species of weevil, *Neochetina eichhorniae* Warner and *N. bruchi* Hustache, a pyralid moth, *Sameodes albiguttalis* Warren and a fungus *Cercospora rodmanii*. The invading submerged aquatics are also often represented by a single genotype, but there are no examples of specific biological control agents known.

It is an extraordinary paradox that the remaining species (58.3%) have very refined outbreeding mechanisms (self-compatibility, tristyly or dioecy) but spread vegetatively and rarely if ever develop seed outside their native range. The pondweed-like genera *Elodea, Hydrilla* and *Lagarosiphon* are dioecious or gynomonoecious and pollinated at or on the surface of the water. The pollination mechanisms are very refined and must represent a considerable investment of resources (Cook, 1982). As the pollen transfer needs water the gametophytic mobility is, at best, confined to a single body of water. These plants also have a very effective (one is tempted to say aggressive) vegetative spread. In nature one rarely finds males and females cohabiting. To date, no allelopathic mechanisms have been discovered but uncritical experiments carried out in small tanks in Zürich indicate that when both sexes of a species are introduced into one tank then within a year one sex is eliminated.

In these invading aquatics it would seem that gene exchange, and thus also the gametophytic mobility, are directly suppressed by aggressive sporophytic growth. The different mating types have developed into competitors. This results in genetic poverty in any particular body of water, to the extent that one sex (or mating type) may eliminate the other. This is particularly extraordinary when the pollination mechanism is geared to pollen transfer within one body of water. The pollen cannot, under normal conditions, travel from one body of water to another. It is difficult to understand how such complex pollination mechanisms can evolve in plants with very efficient clonal growth, or is the lack of sexual reproduction today a result of over specialization of either the pollination mechanisms or subsequent ecological specialization of the sporophytes?

V. Conclusions

Man, through travel and his talent for disturbing the environment has, among other things, provided aquatic plants not only with the means of transport, but also with suitable areas for colonization (artificial lakes, canals, ditches, etc.). Of the 12 most notorious aquatic invaders, only one (*Trapa natans*) is monocarpic and thus relies on sexual processes for its reproduction and spread. All the other species have or can spread by means of vegetative fragments. Only three species regularly develop seed in their native and adventive range (*Myriophyllum spicatum, Najas minor* and *Pistia stratiotes*). The remaining nine species develop no seed in their adventive range. *Salvinia molesta* is a sterile pentaploid and thus denied any sexual reproduction. The rest have well-developed self-incompatibility mechanisms. Genetic incompatibility and very effective vegetative growth lead to a situation where frequently only one breeding type survives in any single body of water.

References

Anderson, L. W. J. (ed.) (1985). Proceedings of the 1st International Symposium on water-milfoil (*Mynophyllum spicatum*) and related Haloragaceae species. Vancouver, 223pp.

Baker, H. B. and Cox, P. A. (1984). Further thoughts on dioecism and islands. *Ann. Mo. Bot. Gard.* **71**, 244–253.

Bawa, K. S. and Beach, J. H. (1981). Evolution of sexual systems in flowering plants. *Ann. Mo. Bot. Gard.* **68**, 254–274.

Cook, C. D. K. (1980). The status of some Indian endemic plants. IUCN. *Threatened Plant Committee Newsletter* **6**, 17–18.

Cook, C. D. K. (1982). Pollination mechanisms in the Hydrocharitaceae. *In* "Studies on Aquatic Vascular Plants" (J.-J. Symoens, S. S. Hooper and P. Compère, eds), pp. 1–15. Royal Bot. Society of Belgium, Brussels.

Cook, C. D. K. (1983). Aquatic plants endemic to Europe and the Mediterranean. *Bot. Jahrb. Syst.* **103**, 539–582.

Cook, C. D. K. (1985). Range extensions of aquatic vascular plant species. *J. Aquat. Plant Management* **23**, 1–6.

Cook, C. D. K. and Lüönd, R. (1982). A revision of the genus *Hydrilla* (Hydrocharitaceae). *Aquat. Bot.* **13**, 485–504.

Cook. C. D. K. and Urmi-König, K. (1984). A revision of the genus *Egeria* (Hydrocharitaceae). *Aquat. Bot.* **19**, 73–96.

Cook, C. D. K. and Urmi-König, K. (1985). A revision of the genus *Elodea* (Hydrocharitaceae). *Aquat. Bot.* **21**, 111–156.

Cook, C. D. K., Gut, B. J., Rix, E. M., Schneller, J. and Seitz, M. (1974). "Water Plants of the World". Junk, The Hague.

Givnish, T. J. (1980). Ecological constraints on the evolution of breeding systems in plants: dioecy and dispersal in gymnosperms. *Evolution,* **34**, 959–972.

Gopal, B. and Sharma, K. P. (1981). "Water Hyacinth, the Most Troublesome Weed of the World". Hindasia, Delhi.

Jain, S. K. (1976). The evolution of inbreeding in plants. *Ann. Rev. Ecol. Syst.* **7**, 469–495.

Loyal, D. S. and Grewal, R. K. (1966). Cytological study on sterility in *Salvinia auriculata* Aublet with a bearing on its reproductive mechanism. *Cytologia* **31**, 330–338.

Room, P. M. (1986). Biological control is solving the World's *Salvinia molesta* problems. Proceedings of the EWRS/AAB 7th Symposium on Aquatic Weeds, Loughborough, pp. 271–276.

Schneller, J. J. (1981). Chromosome number and spores of *Salvinia auriculata* Aublet s. str. *Aquat. Bot.* **10**, 81–84.

Sculthorpe, C. D. (1967). "The Biology of Aquatic Vascular Plants". Edward Arnold, London.

Sobrevila, C. and Kalin Arroyo, M. T. (1982). Breeding systems in a montane tropical cloud forest in Venezuela. *Plant Syst. Evol.* **140**, 91–138.

Willson, M. F. (1979). Sexual selection in plants. *Am. Nat.* **113**, 777–790.

12 Differentiation Trends in Tropical Woody Angiosperms

FRIEDRICH EHRENDORFER

Institute of Botany, University of Vienna, Vienna, Austria

I. Introduction

In this chapter I would like to present the picture emerging from more than 20 years of research work on the karyosystematics and evolution of tropical Angiosperm families, carried out by our team at the Vienna Institute of Botany. Obviously, it will not be possible to bring forward within the framework of this contribution all of the underlying factual data. Therefore, I will only briefly discuss examples from several tropical woody and some related temperate Angiosperm families. Trends in the variation and differentiation of these groups include:

(1) life and growth forms, vegetative adaptations and ecological preferences;
(2) flower and fruit biology, reproductive systems, and dispersal strategies;
(3) phytochemical constitution, accumulation of secondary compounds and defence mechanisms against herbivores and parasites;
(4) chromosome numbers, karyotypes, structures of interphase nuclei, eu- and heterochromatin, genome size and DNA quantities;
(5) spatial organization and interrelationships of taxa, isolation and hybridization, size and stability of populations.

For more detail, the reader is referred to the extensive literature sources (more general surveys are found in Ehrendorfer *et al.*, 1968; Ehrendorfer, 1976, 1980, 1982, 1986, and in press; and Morawetz, 1986a).

Differentiation Patterns in Higher Plants
ISBN: 0-12-709425-3

II. Case Studies

A. Magnoliidae: Winteraceae, Annonaceae, Laurales, Aristolochiaceae

The Winteraceae are a very isolated relic and small family of the Magnoliidae (about 50 species and 4–8 genera: Vink, 1970, 1985; Ehrendorfer *et al.*, 1979), exhibiting numerous primitive morphological and anatomical features. They mostly occur in humid to mesic (sub)tropical-montane habitats. At present, the Winteraceae have a strongly disjunct distribution, centred in the Western Pacific area, but extending to Indomalaysia, East Australia and Tasmania, New Zealand, South and Central America, and to Madagascar. Formerly, their distribution area was much more extensive, dating back to the Upper Cretaceous (Coetzee and Muller, 1984). With the exception of the dioecious genus *Tasmannia* with $x = 13$, all the other genera studied so far as highly paleopolyploid on $x = 43$, with $2n = 86$, and occasionally even $2n = 172$ (Ehrendorfer *et al.*, 1968). The karyotype with $x = 43$ is strongly differentiated and diploidized (Morawetz, 1984a). The rather generalized flowers are pollinated by a variety of insects. *Drimys* is self-fertile (Gottsberger *et al.*, 1980), and *Pseudowintera* self-sterile (Godley and Smith, 1981). While most of the genera have apparently maintained little potential for further differentiation and adaptation, *Tasmannia* in New Guinea (Vink, 1970), and particularly *Drimys* in the New World (Ehrendorfer *et al.*, 1979), exhibit remarkable eco-geographical radiation with partial or complete extrinsic isolation of races with different growth forms in different habitats or with different flowering periods.

With more than 2000 species the Annonaceae are a large pantropical family. The available systematic arrangement of the *c.* 130 genera (Fries, 1959) follows formal morphological lines and often does not reflect their natural affinities. This is clearly demonstrated by an improved understanding of palynological features, both on the LM and the TM level (e.g. Walker, 1971; Le Thomas, 1980/81), and the increasing amount of available karyological (e.g. Ehrendorfer *et al.*, 1968; Tanaka and Okada, 1972; Okada and Ueda, 1984; Sauer and Ehrendorfer, 1984; Morawetz, 1986a) and phytochemical information (e.g. Leboeuf *et al.*, 1982). An example for the clarification of natural relationships is afforded by the interesting genus *Anaxagorea*, which has recently been studied with more intensity (Maas and Westra, 1984/85; Hesse *et al.*, 1985; Le Thomas, 1984). The majority of its species are neotropical, but there are two independent links with the three species in southeast Asia. On the generic level *Anaxagorea* exhibits affinities with several West African genera, suggesting an ancient transatlantic Western Gondwanaland distribution.

Secondary phytochemical constituents are very diverse within the Annonaceae, but more data will have to be accumulated in order to demonstrate better generic differences and ecological significance. Reproductive biology is often quite specialized in the family (e.g. in *Guatteria*: pollination by fruit-visiting beetles, breeding in the long-lasting flowers; Gottsberger, 1970).

On the whole, chromosomal evolution has been slow relative to morphological diversification. However, there are instances of infraspecific chromosome structural differentiation (e.g. in *Porcelia goyazensis*: Morawetz, 1984a) or infraspecific polyploidy (including triploids which apparently are agamospermous, e.g. in *Duguetia furfuracea*: Morawetz, 1984b). Such cases appear more exceptional compared with genera like *Annona*, where an extraordinary species diversity has originated without any obvious chromosomal changes. This is clearly brought out by the application of modern Giemsa C- and fluorochrome banding methods which demonstrate near-identical patterns even in species of different sections (Morawetz, 1984c, 1986a). Thus, *Annona* in the New World has successfully radiated from humid rain forests into dry and fire-influenced savanna or cerrado habitats, producing several more and more reduced pyrophytes. The occasional establishment of polyploid species in *Annona* obviously contributes as an isolating mechanism between different sympatric species.

In contrast, different genera of Annonaceae often exhibit different chromosome banding patterns and sometimes even different base numbers, such as $x = 7$, 8, and 9. As an example one can point to the *Annona* subfamily where the *Hexalobus* and the *Asimina* groups of genera have $x = 8$, the *Annona* group $x = 7$ and the *Cymbopetalum* group $x = 9$ (Morawetz, 1986a; Ehrendorfer, 1986, and in press).

In several Annonaceae one can demonstrate the successive shift of kladogenetic differentiation from the diploid to polyploid levels: there are neopolyploids in *Duguetia* or *Cymbopetalum*, and well-stabilized mesopolyploid species, or even sections in *Annona* (e.g. *A. glabra*, $4x$; *A. pygmaea*, $4x$ and *A. coriacea*, $6x$). Finally, there are groups of genera, such as the *Guatteria* tribe (Morawetz and Waha, 1985), which have lost all their diploid progenitors and have continued to differentiate as paleopolyploids on a clearly diploidized and remarkably stabilized karyotype of $2n = 28$ (probably $4x$ on $x = 7$).

While there is certainly some differentiation of chromosome size, DNA quantity and interphase nuclear structure within Annonaceae (cf. *Tetrameranthus*: Morawetz, 1986b), such nucleotypic diversification has become spectacular within the Laurales families (Morawetz, 1986a). Within the predominantly diploid and very relictual Trimeniaceae and Chloranthaceae (with $2n = 16$, etc.) we find very large chromosomes and densely euchromatic interphase nuclei. In contrast, the less depauperate and highly specialized (flower structures and reproductive biology: Endress, 1980; Lorence, 1985) Monimiaceae s.str. have evolved very small chromosomes on a highly paleopolyploid level ($2n = 38-44$) and lightly staining prochromosomal interphase nuclei. Other families (e.g. the Lauraceae, Gyrocarpaceae, Hernandiaceae, Siparunaceae, or Atherospermataceae) maintain a somewhat intermediate position between these two extremes. Within the Monimiaceae, some genera like *Hedycarya* or *Mollinedia* are in the process of producing another cycle or neopolyploids and dysploids ($2n = 36$, 38, 40, 100, 114, 180), while others, such as the relictual and monotypic Chilean *Peumus*, have long since become stabilized on $2n = 78$ and exhibit a strictly diploidized and highly differentiated karyotype (Morawetz, 1981).

It is fascinating to compare the relatively primitive woody Annonaceae (or Laurales) with a closely related but much more advanced family such as the Aristolochiaceae (about 600 species in 6–14 genera: Huber, 1985; Morawetz, 1985; Sugawara, 1981, 1982, 1985a,b; Sugawara and Ogisu, 1986). This family has evolved into a great variety of different habitats and ranges from tropical to temperate zones. Parallel to this, highly advanced liana as well as herbaceous life forms have originated. Furthermore, flower morphology has become more and more extremely specialized. Phytochemical constituents are less diverse and partly more advanced. In comparison with woody Magnoliidae, the evolution of the Aristolochiaceae has been accompanied by a much larger degree of karyotype and nucleotype differentiation, as is documented by the variety of different base numbers and banding patterns.

B. Rosidae and Dilleniidae: Rutaceae, Dilleniaceae, Paeoniaceae

The Rutaceae are a large family with *c.* 150 genera and about 1800 species which have radiated widely from humid and mesic into dry and xeric habitats, and from the tropics into the subtropical and temperate zones. Here again, the last complete systematic treatment (Engler, 1931) proves to be largely formal and "artificial" on the basis of more recent studies, particularly phytochemical (Fish and Waterman, 1973; Waterman, 1975; Waterman and Grundon, 1983; Fernandes da Silva *et al.*, in prep.) and karyological (Guerra, 1980, 1984a,b, and in press; Ehrendorfer, 1982). As an illustration one may point to the genera *Zanthoxylum* and *Toddalia* which have been placed into different subfamilies but prove to be very close in their characteristic benzylisochinoline alkaloids, their paleopolyploidy and their similar interphase nuclei.

Rutaceae are predominantly woody; nevertheless, a great variety of life forms and leaf types has evolved, all of a clearly adaptive nature. Parallel with this radiation, a shift in the accumulation of secondary compounds has taken place. In a few of the most primitive groups benzylisochinoline alkaloids (otherwise mostly found in Magnoliidae only) have been preserved. In more advanced groups these compounds have been replaced by simple and later by more complex anthranilic acid derivatives and these, in turn, by prenylated coumarins in the most progressive groups.

It is remarkable that these changes in phytochemical defence strategies are paralleled by different phases of chromosomal evolution in the Rutaceae. In the large pantropical genus *Zanthoxylum*, as well as in *Toddalia* and *Phellodendron*, all with benzylisochinoline alkaloids and dating back into the late Cretaceous or early Tertiary, one encounters exclusively paleopolyploids and related dysploids. In contrast, the Aurantioideae (with the genus *Citrus* and its allies) are much younger and have radiated mostly in humid and mesic habitats from a southeast

Asiatic centre. They are much more advanced, both in their morphology and phytochemistry, have remained almost exclusively on the diploid chromosome level, and do not exhibit any major differentiation of chromosome base number, karyotype and/or interphase nuclear structure.

Another and even more progressive phase of chromosome evolution has paralleled the "explosive" radiation of the family into the dry and nutrient-deficient habitats of the sclerophyllous floras of the Cape and southeast/southwest Australia. The resulting two tribes, the Diosmeae in South Africa and the Boronieae in Australia, both have numerous and closely linked genera and species; both clearly had an independent and parallel origin from tropical-montane members of the *Evodia* alliance which today is paleopolyploid and rather uniform in its chromosome structures and numbers. This contrasts very sharply with the chromosomal situation in the Diosmeae and Boronieae, which are both characterized by an enormous array of different dysploid and polyploid chromosome numbers, reflecting their intensive kladogenetic differentiation. At the same time the two tribes are amongst the most progressive with respect to their phytochemistry (e.g. Leboeuf *et al.*, 1982).

A comparison of the DNA values and the evolutionary potentials in the different groups of Rutaceae suggests that both relatively low and very high DNA quantities have a stabilizing effect, whereas medium quantities tend to stimulate chromosomal differentiation and dysploidy (Ehrendorfer, 1982; Guerra, 1980, 1984b).

With 11 genera and about 325 species, the Dilleniaceae are one of the medium-sized and better understood pantropical families of Angiosperms (see e.g. Gilg and Werdermann, 1925; Kubitzki, 1975; Dickison *et al.*, 1982; Ehrendorfer, 1980, 1982, in press, and unpubl.).

There are two subfamilies, the more basic Dillenioideae centred in Australasia and extending to Madagascar, and the more advanced Tetraceroideae centred in the Neotropics, but also in Africa and southeast Asia. The chromosome base number of the family appears to be $x = 7$; this number has survived only in paleopolyploid members from mesic forest habitats of the Paleotropics and is found in both subfamilies. It is remarkable that the greatest conservation of character states considered "primitive" with regard to morphology, anatomy, phytochemistry, etc. is also found among these paleopolyploid representatives.

Within the Dillenioideae, the genus *Dillenia* (with c. 65 species from Australasia to Madagascar) takes a central position. It includes rain forest trees (rarely shrubs) with unspecialized flowers which are $8x$ ($n = 27, 26, 24$) and $16x$ ($n = 56$). Also limited to mesic rain forest habitats are the woody genera *Didesmandra* (monotypic and endemic in Borneo, $6x$, $n = 21$) and *Schumacheria* (with a few species endemic in Sri Lanka, $8x$, $n = 28$); $8x$, $n = 28$ also has been recorded for *Acrotrema*. The *c.* 10 species of this genus have switched to a herbaceous life form and are colonizing rocky outcrops in rain forest areas from Sri Lanka to southern India and from the Malay Peninsula to southern Burma. In contrast to these rather depauperate and paleopolyploid genera of Dillenioideae,

Hibbertia has diversified extensively with the majority of its *c.* 150 species in Australia (Stebbins and Hoogland, 1976; Rury and Dickison, 1977; Dickison *et al.*, 1978; Ehrendorfer, 1980, 1982, and in press). Its morphologically most primitive tree-like taxa occupy rather mesic habitats and belong to the extremely ancient relic flora of New Caledonia; again, they are paleopolyploids on $8x$ (with $n = 27, 28$). In Australia, an explosive ecological and geographical radiation has produced advanced growth forms, ranging from small trees of mesic habitats to climbers or undershrubs of dry woodlands, and lastly to pygmy xerophytes of semi-deserts. This has been accompanied by the development of an astonishing array of different chromosome structures and numbers ($n = 4, 5, 8, 9, 10, 12, 16, 18, 32$) which can be interpreted as a combination of descending dysploidy and ascending polyploidy. Linked to *Hibbertia*, there is also the juncoid and practically leafless genus *Pachynema* (*c.* 3 species, one with $n = 13$) from the semi-deserts of central Australia.

The Tetraceroideae (with about 100 species) have diversified on a paleo-$4x$ level and descending dysploidy. They include the pantropical *Tetracera* ($n = 14, 13$), which apparently originated in Africa, and the neotropical genera *Doliocarpus, Curatella* (both with $n = 13$), *Davila* ($n = 10$), and *Pinzona*. The evolution of the Tetraceroideae is characterized by their radiation from mesic rain forests into dry and open woodlands or even savannas and cerrados.

Thus, within Dilleniaceae, we can recognize three major evolutionary phases: the $6x$-, $8x$- and $16x$-paleopolyploid Dillenioideae, often depauperate and local endemics of ancient stable rain forest habitats, are apparently relics from an early differentiation phase of the family. A second phase is represented by the Tetraceroideae, differentiated on a $4x$-paleopolyploid level and relatively well preserved in the drier tropical habitats of Africa and the New World. Finally, the Australian species of *Hibbertia* and *Pachynema* obviously correspond to a third and more recent phase. Here, rapid diversification and radiation into a multitude of more or less dry to xeric, nutrient-deficient, and floristically unsaturated habitats has been paralleled by the origin of an extensive dysploid and diploid–polyploid chromosome pattern (comparable to Rutaceae–Boronieae and –Diosmeae!).

The monogeneric Paeoniaceae (with *c.* 50 species; Stern, 1946; Stebbins, 1948) are related to the Dilleniaceae and can be regarded as its northern hemisphere temperate derivative. *Paeonia* has advanced from a shrubby to a perennial herbaceous growth form and today appears in three major disjunct distribution areas, in east Asia, in Europe, the Mediterranean and southwest Asia, and in western North America. Advanced evolutionary phenomena are extensive neopolyploidy in the Eurasiatic species ($n = 5, 10$) and balanced translocation heterozygosity in the North American species (Walters, 1942).

C. Asteridae: Rubiaceae

With more than 6000 species and about 500 genera the Rubiaceae are among the largest and most successful families of Angiosperms, radiating from the tropics to the temperate and even the arctic/antarctic zones. Only recent multi-disciplinary systematic studies, covering aspects of karyology (e.g. Ehrendorfer, 1977, and in press; Kiehn, 1985, 1986a,b, and in press), phytochemistry (e.g. Kisakürek *et al.*, 1983), seed anatomy (e.g. Lorenz-Liburnau, 1986) etc., are gradually furnishing the basis for a more natural arrangement of the family. As an example one can point to the heterogeneous Cinchoneae s. lat. where the modern methods listed above have helped to segregate several truly related generic assemblies, e.g. the neotropical and paleopolyploid groups with *Cinchona, Remijia*, etc. ($n = 17$) and with *Posoqueria* ($n = 17-18$), the paleotropical group with *Hymenodictyon*, etc. ($n = 11$, $2x-4x-6x$), and the predominantly paleotropical Mitragyninae with *Halea, Mitragyna* and *Uncaria* (all paleo-$4x$, $n = 22$), also closely linked to the paleotropical and paleo-$4x$ Naucleeae (Ridsdale, 1978; Phillipson *et al.*, 1982).

Within the large family there is a remarkable change in the accumulation tendencies of different phytochemical compounds, with indole alkaloids occurring mainly in the Cinchonoideae, and with asperulosid glycosides and antrachinones occurring predominantly in the Rubioideae. Other main lines of evolutionary differentiation concern the flower and fruit morphology, clearly linked to different pollination strategies (zoophily towards anemophily, homostyly towards heterostyly and dioecy, allogamy towards autogamy, etc.). As an example one can list the Anthospermeae (Puff, 1986), which have switched to anemophily and have mostly become pioneer plants of more or less open habitats in a great variety of climatic zones. They can be segregated into the Anthosperminae, centred in Africa, the amphipacific Coprosminae, and the Australian Operculariinae. The African Anthosperminae have clearly radiated from humid or mesic forest to more open and eventually xeric habitats, and demonstrate interesting parallel changes in growth form (from small trees to annual herbs) and leaf anatomy.

Within the Rubiaceae different phases of evolutionary differentiation are clearly apparent. Many of the morphologically, anatomically, and phytochemically "primitive" groups have only survived as paleopolyploids, e.g. most of the Cinchoneae, the Naucleeae, Guettardeae, etc. Other woody groups at a medium level of advancement still retain some of their diploids or are still represented predominantly on the diploid level; their chromosome base numbers are apparently mostly stabilized, for example in the Coffeeae, Gardenieae, Psychotrieae, etc. In contrast, several of the most progressive tribes which have changed from woody to herbaceous and even annual growth forms, exhibit a great variety of different dysploids and additional polyploids, e.g. the Hedyotideae ($x = 11, 10, 9, 8, 7, 6$), the Paederieae ($x = 13, 12, 11, 10$), or the Rubieae ($x = 12, 11, 10, 9, 5$).

III. Conclusions

I would like to stress again how incomplete and preliminary much of our knowledge about tropical woody Angiosperms still is. Obviously, the more general conclusions about differentiation trends presented here in a comparison with temperate herbaceous groups will have to be tentative and hypothetical. On the other hand, certain correlations among the trends discussed can be characterized as syndromes of evolutionary differentiation, the features listed below being involved.

(1) Changes in *growth form*, leaf anatomy, etc. linked to *eco-radiation* from tropical rain forest to dry and secondary bush, tropical savanna, mediterranoid sclerophyll and temperate habitats.
(2) More diversity and eco-radiation into specialized *flower–pollinator* and *fruit/seed–animal consumer/disperser* relationships.
(3) More diversity in the accumulation of secondary compounds; changes in *phytochemical defence strategies* often correlated with eco-radiation into different habitats and with successive phases of phylogenetic differentiation.
(4) Comparable spectrum of *karyotypic and nucleotypic differentiation*; relative to structural and functional differentiation mostly slow and "conservative" in stable rain forest habitats, increasingly accelerated and "progressive" in more labile and drier habitats.
(5) *Populations* of tropical forest species often more dispersed, with more extrinsic *isolation*, stepped clinal variation and less hybridization; this trend is reversed in more labile and open habitats.
(6) Within tropical families broader and better conserved spectrum of representatives from *successive phases of phylogenetic differentiation*: from ancient and fractionated, stasigenetic groups, paleopolyploids and relics from early anagenetic radiation, with a high proportion of "primitive" features, often in ancient and stable, humid habitats, to more recent and coherent groups, diploids or dysploids and neopolyploids in full kladogenetic radiation, with many "advanced" features, often in more recent, labile and drier habitats.

I believe that it will be both stimulating and rewarding to continue with this type of inductive and synthetic research, and to intensify it through international cooperation.

Acknowledgements

For support and help I am grateful to all collaborators of our tropical botany research team in Vienna and abroad. The "Fonds zur Förderung der wissenschaftlichen Forschung in Oesterreich" has financed much of our research (Projects 4052 and 4799).

References

Coetzee, J. A. and Muller, J. (1984). The phytogeographic significance of some extinct Gondwana pollen types from the Tertiary of the Southwestern Cape (South Africa). *Ann. Mo. Bot. Gard.* **71**, 1088–1099.

Dickison, W. C., Rury, P. M. and Stebbins, G. L. (1978). Xylem anatomy of *Hibbertia* (Dilleniaceae) in relation to ecology and evolution. *J. Arnold Arbor* **59**, 32–49.

Dickison, W. C., Nowicke, J. W. and Skvarla, J. J. (1982). Pollen morphology of the Dilleniaceae and Actinidiaceae. *Am. J. Bot.* **69**, 1055–1073.

Ehrendorfer, F. (1976). Evolutionary significance of chromosomal differentiation patterns in Gymnosperms and primitive Angiosperms. *In* "Origin and Early Evolution of Angiosperms" (C. B. Beck, ed.), pp. 220–240. Columbia University Press, New York.

Ehrendorfer, F. (1977). Karyologie, Systematik und Evolution der Rubiaceae. *Symp. Morph. Anat. Syst. Pfl.*, Strasbourg, 1977.

Ehrendorfer, F. (1980). Polyploidy and distribution. *In* "Polyploidy, Biological Relevance" (W. Lewis, ed.), pp. 45–60. Plenum, New York.

Ehrendorfer, F. (1982). Speciation patterns in woody Angiosperms of tropical origin. *In* "Mechanisms of Speciation" (C. Barigozzi, ed.), pp. 479–509. A. R. Liss, New York.

Ehrendorfer, F. (1986). Chromosomal differentiation and evolution in Angiosperm groups. *In* "Modern Aspects of Species" (K. Iwatsuki, P. H. Raven and W. J. Bock, eds), pp. 59–86. University of Tokyo Press, Tokyo.

Ehrendorfer, F. (In press). Affinities of the African dendroflora: suggestions from karyo- and chemosystematics. *Ann. Mo. Bot. Gard.*

Ehrendorfer, F., Krendl, F., Habeler, E. and Sauer, W. (1968). Chromosome numbers and evolution in primitive Angiosperms. *Taxon* **17**, 337–353.

Ehrendorfer, F., Silberbauer-Gottsberger, I. and Gottsberger, G. (1979). Variation on the population, racial and species level in the primitive relic Angiosperm genus *Drimys* (Winteraceae) in South America. *Plant Syst. Evol.* **132**, 53–83.

Endress, P. (1980). Ontogeny, function and evolution of extreme floral construction in Monimiaceae. *Plant Syst. Evol.* **134**, 79–120.

Engler, A. (1931). Rutaceae. *In* "Die natürlichen Pflanzenfamilien" (A. Engler and K. Prantl, eds), 2nd Ed., pp. 187–359, 458–459. W. Engelmann, Leipzig.

Fernandes da Silva, M. F., Gottlieb, O. R. and Ehrendorfer, F. (in prep.). Major alkaloids and coumarins in Rutaceae: suggestions for a natural system and evolutionary interpretation of the family.

Fish, F. and Waterman, P. G. (1973). Chemosystematics in the Rutaceae. II. The chemosystematics of the *Zanthoxylum/Fagara* complex. *Taxon* **22**, 177–203.

Fries, R. E. (1959). Annonaceae. *In* "Die natürlichen Pflanzenfamilien" (A. Engler and K. Prantl, eds), 2nd Ed., pp. 1–171. Duncker and Humblot, Berlin.

Gilg, E. and Werdermann, E. (1925). Dilleniaceae. *In* "Die natürlichen Pflanzenfamilien" (A. Engler and K. Prantl, eds), 2nd Ed., pp. 7–47. W. Engelmann, Leipzig.

Godley, E. J. and Smith, D. H. (1981). Breeding systems in New Zealand plants. 5. *Pseudowintera colorata* (Winteraceae). *N. Zealand J. Bot.* **19**, 151–156.

Gottsberger, G. (1970). Beiträge zur Biologie von Annonaceen-Blüten. *Oesterr. Bot. Z.* **118**, 237–279.

Gottsberger, G., Silverbauer-Gottsberger, I. and Ehrendorfer, F. (1980). Reproductive biology in the primitive relic Angiosperm *Drimys brasiliensis* (Winteraceae). *Plant Syst. Evol.* **135**, 11–39.

Guerra, M. dos S. (1980). Karyosystematik und Evolution der Rutaceae. Diss. Formal- u. Naturwiss. Fak. Univ. Vienna.

Guerra, M. dos S. (1984a). New chromosome numbers in Rutaceae. *Plant Syst. Evol.* **146**, 13–30.
Guerra, M. dos S. (1984b). Cytogenetics of Rutaceae. II. Nuclear DNA content. *Caryologia* **37**, 219–226.
Guerra, M. dos S. (In press). Cytogenetics of Rutaceae. IV. Structure and systematic significance of interphase nuclei. *Cytologia* **52**.
Hesse, M., Morawetz, W. and Ehrendorfer, F. (1985). Pollen ultrastructure and systematic affinities of *Anaxagorea* (Annonaceae). *Plant Syst. Evol.* **149**, 149–153.
Huber, H. (1985). Samenmerkmale und Gliederung der Aristolochiaceen. *Bot. Jb. Syst.* **107**, 277–320.
Kiehn, M. (1985). Karyosystematische Untersuchungen an Rubiaceae: Chromosomenzählungen aus Afrika, Madagaskar und Mauritius. *Plant Syst. Evol.* **149**, 89–118.
Kiehn, M. (1986a). Karyologische Untersuchungen und DNA-Messungen an Rubiaceae und ihre Bedeutung für die Systematik dieser Familie. Diss. Formal- u. Naturwiss. Fak. Univ., Vienna.
Kiehn, M. (1986b). Karyosystematic studies on Rubiaceae: Chromosome counts from Sri Lanka. *Plant Syst. Evol.* **154**, 213–223.
Kiehn, M. (In press). Karyosystematische Untersuchungen an Rubiaceae: Chromosomenzählungen aus den Neotropen. *Flora*.
Kisakürek, M. V., Leeuwenberg, A. J. M. and Hesse, M. (1983). A chemotaxonomic investigation of the plant families of Apocynaceae, Loganiaceae and Rubiaceae by their indole alkaloid content. *In* "Alkaloids: Chemical and Biological Perspectives", Vol. 1 (S. W. Pelletier, ed.), pp. 211–376. John Wiley & Sons, New York.
Kubitzki, K. (1975). Relationships between distribution and evolution in some heterobathmic tropical groups. *Bot. Jb. Syst.* **96**, 212–230.
Leboeuf, M., Cave, A., Bhaumik, P. K., Mukherjee, B. and Mukherjee, R. (1982). The phytochemistry of the Annonaceae. *Phytochemistry* **21**, 2783–2813.
Le Thomas, A. (1980/81). Ultrastructural characters of the pollen grains of African Annonaceae and their significance for the phylogeny of primitive Angiosperms, I. + II. *Pollen and Spores* **22**, 267–345; **23**, 5–36.
Le Thomas, A. (1984). Evidences palynologiques sur l'origine des Annonaceae (Magnoliales). *Rev. Paléobiol. Vol. Spec.*, 119–124.
Lorence, D. H. (1985). A monograph of the Monimiaceae (Laurales) in the Malagasy region (Southwest Indian Ocean). *Ann. Mo. Bot. Gard.* **72**, 1–165.
Lorenz-Liburnau, E. (1986). Samenanatomische Merkmale bei verschiedenen Rubiaceen — ein Beitrag zur Systematik der Familie. Diss. Formal- u. Naturwiss. Fak. Univ. Vienna.
Maas, P. J. M. and Westra, L. Y. (eds) (1984/5). Studies in Annonaceae. II. A monograph of the genus *Anaxagorea* A. St. Hil. H, 1+2. *Bot. Jb.* **105**, 73–134; 145–204.
Morawetz, W. (1981). Karyologie und ökologisch-morphologische Differenzierung von *Peumus boldus* (Monimiaceae, Laurales). *Plant Syst. Evol.* **138**, 157–173.
Morawetz, W. (1984a). How stable are genomes of tropical plants? Heterozygosity of C-banded karyotypes of *Porcelia* as compared with *Annona* (Annonaceae) and *Drimys* (Winteraceae). *Plant Syst. Evol.* **145**, 29–39.
Morawetz, W. (1984b). Karyological races and ecology of Brazilian *Duguetia furfuracea* as compared with *Xylopia aromatica* (Annonaceae). *Flora* **175**, 195–209.
Morawetz, W. (1984c). Karyologie, Oekologie und Evolution der Gattung *Annona* (Annonaceae) in Pernambuco, Brasilien. *Flora* **175**, 435–447.
Morawetz, W. (1985). Beiträge zur Karyologie und Systematik der Gattung *Thottea* (Aristolochiaceae). *Bot. Jb. Syst.* **107**, 329–342.
Morawetz, W. (1986a). Remarks on karyological differentiation patterns in tropical woody plants. *Plant Syst. Evol.* **152**, 49–100.

Morawetz, W. (1986b). Systematics and karyoevolution in Magnoliidae: *Tetrameranthus* as compared with other Annonaceous genera of the same chromosome number. *Plant Syst. Evol.* **154**, 147–173.

Morawetz, W. and Waha, M. (1985). A new pollen type, C-banded and fluorescent counterstained chromosomes, and evolution in *Guatteria* and related genera (Annonaceae). *Plant Syst. Evol.* **150**, 119–141.

Okada, H. and Ueda, K. (1984). Cytotaxonomical studies on Asian Annonaceae. *Plant Syst. Evol.* **144**, 165–177.

Phillipson, J. D., Hemingway, S. R. and Ridsdale, C. E. (1982). The chemotaxonomic significance of alkaloids in the Naucleeae s.l. (Rubiaceae). *J. Nat. Prod.* **45**, 115–162.

Puff, Ch. (1986). A biosystematic study of the African and Madagascan Rubiaceae-Anthospermeae. *Plant Syst. Evol.*, Suppl. 3. Springer, Vienna and New York.

Ridsdale, C. E. (1978). A revision of the tribe Naucleeae s.s. (Rubiaceae). *Blumea* **24**, 43–100.

Rury, P. M. and Dickison, W. C. (1977). Leaf venation patterns of the genus *Hibbertia* (Dilleniaceae). *J. Arnold Arbor.* **58**, 209–241.

Sauer, W. and Ehrendorfer, F. (1984). Notes on the karyosystematics of Annonaceae. *Plant Syst. Evol.* **146**, 47–55.

Stebbins, G. L. (1948). Review of "A Study of the Genus *Paeonia*" by F. C. Stern. *Madroño* **9**, 193–199.

Stebbins, G. L. and Hoogland, R. D. (1976). Species diversity, ecology and evolution in a primitive Angiosperm genus: *Hibbertia* (Dilleniaceae). *Plant Syst. Evol.* **125**, 139–154.

Stern, F. C. (1946). "A Study of the Genus *Paeonia*". Royal Horticultural Society, London.

Sugawara, T. (1981). Taxonomic studies of *Asarum* sensu lato. I. Karyotype and C-banded pattern in *Asarum* s. str., *Asiasarum* and *Heterotropa*. *Bot. Mag. Tokyo* **94**, 225–238.

Sugawara, T. (1982). Taxonomic studies of *Asarum* sensu lato. II. Karyotype and C-banded pattern in two species of *Hexastylis* and *Asarum epigynum*. *Bot. Mag. Tokyo* **95**, 295–302.

Sugawara, T. (1985a). Further investigations on karyotype and C-banded pattern in seven species of *Heterotropa*. *J. Jap. Bot.* **60**, 15–24.

Sugawara, T. (1985b). A note on *Heterotropa pseudosavatieri* and its allied species (Aristolochiaceae). *J. Jap. Bot.* **60**, 202–212.

Sugawara, T. and Ogisu, M. (1986). Karyotype analysis of five species of *Asarum* (Aristolochiaceae) in Sichuan, China. *J. Jap. Bot.* **62**, 104–111.

Tanaka, R. and Okada, H. (1972). Karyological studies in four species of Annonaceae, a primitive Angiosperm. *J. Sc. Hiroshima Univ.*, Ser. B/2 (Bot.) **14**, 85–105.

Vink, W. (1970). The Winteraceae of the Old World. I. *Pseudowintera* and *Drimys*, morphology and taxonomy. *Blumea* **18**, 225–354.

Vink, W. (1985). The Winteraceae of the Old World. V. *Exospermum* links *Bubbia* to *Zygogynum*. *Blumea* **31**, 39–55.

Walker, W. J. (1971). Pollen morphology, phytogeography and phylogeny of the Annonaceae. *Contr. Gray Herb.* **202**, 1–132.

Walters, J. L. (1942). Distribution of structural hybrids in *Paeonia californica*. *Am. J. Bot.* **29**, 270–275.

Waterman, P. G. (1975). Alkaloids of the Rutaceae: their distribution and systematic significance. *Biochem. Syst.* **3**, 149–180.

Waterman, P. G. and Grundon, M. F. (eds) (1983). "Chemistry and Chemical Taxonomy of the Rutales". Academic Press, London and New York.

13 Breeding Systems in the Vascular Flora of Arctic North America

DAVID F. MURRAY

University of Alaska Museum, Fairbanks, Alaska, U.S.A.

I. Introduction

I have been stimulated to make this review by my unwillingness to accept the long held and often repeated view that vascular plants in the Arctic (and by extension also in alpine tundra) have to a large degree substituted asexual methods of reproduction for sexual means. The case for this belief made by Löve (1959) was subsequently incorporated into reviews by several others whose experience, especially in the widespread, sedge- and grass-dominated wet meadows of the arctic coastal plain of Alaska, provided ample evidence of a vegetation type composed of species with extremely vigorous vegetative growth, with infrequent flowering relative to biomass present, and with few or no seedlings as evidence of reproductive success (e.g. Billings and Mooney, 1968; Johnson, 1969; Bliss, 1971; Johnson and Tieszen, 1973; Billings, 1974). There is also the suggestion that it is, energetically, far less expensive to produce tillers and rhizomes than inflorescences and seeds (Chapin *et al.*, 1980). Given, therefore, the energy limitations at high latitude, a trend away from the expensive sexual process made perfect sense.

The assertion that ''bulbil formation is very frequent, and nowhere else is there a higher frequency of asexual seed formation, or apomixis'' (Löve, 1959) is less well supported by direct observations from the American Arctic. Bulbils are formed by *Polygonum viviparum, Saxifraga cernua, S. foliolosa*, and *S. flagellaris*. Except for *S. cernua*, which has rarely been seen to produce seeds, each of the species mentioned above is also capable of seed production. Additionally there is the

Differentiation Patterns in Higher Plants
ISBN: 0-12-709425-3

viviparous habit of some grasses (e.g. *Poa alpina, P. arctica, P. alpigena*, and *Festuca vivipara*) as an example of specialization for vegetative reproduction. Species of the highly apomictic genera *Antennaria* and *Taraxacum* are several and widespread. Yet all of these reproductive specialists together hardly constitute a shift to asexual reproduction for the vascular flora. Nevertheless, the structural simplicity and lower energetic demand give to the asexual process a greater assurance of success in a region where short, cold summers could in many years preclude reproduction by less streamlined methods. It has been reasoned that such an environment would select for the vegetative and apomictic forms, and it is within this limited framework of adaptation that the reproductive behaviour has been analysed and interpreted. The asexual modes of reproduction are treated as adaptations specific to arctic conditions. Yet my own reaction as a taxonomist working with this flora has been that taxa never producing seed are very rare and that taxa producing seeds copiously are common, even among those with highly developed means for rapid vegetative ramification.

II. Arctic Environment

In my view, part of the interpretive difficulty lies in the very generalized view of the arctic environment as *uniformly* and *predictably harsh*. That harshness, although quite real, tends to be translated into a singular environmental force that directs and drives natural selection. A singular response guaranteed by genotypic conservatism is presumably adaptive, hence favoured. But is that true? What really is the nature of the arctic environment to which plant life there must be adapted?

The Arctic is the area of the world north of latitudinal tree-line. In relative terms, it is an energy-limited ecosystem. The low solar angle at high latitudes determines a longer path that solar radiation must travel through the atmosphere, allowing for a greater loss of energy through absorption and scatter before it reaches the surface. Therefore sunlight in the north is always received at lower intensities than to the south. Total irradiance, however, is comparable in the Arctic to that at much lower latitudes due to the essentially continuous daylight during the growing season. Pervasive cloud cover and fog wherever there is open water can profoundly affect net radiation regionally. While adjustments of the photosynthetic apparatus and acclimation are effective adaptations to this light regime, low temperature remains a significant limitation to the distribution of plants in the Arctic. A progressive diminution of net solar radiation (and thus also summer warmth) northward is expressed in the treelessness of tundra and in the vegetational and floristic zonation evident there (Polunin, 1951; Young, 1971; Hare and Ritchie, 1972; Alexandrova, 1977; Yurtsev *et al.*, 1978; Bliss, 1981). There is now a rich literature accruing from ecological investigations conducted under the auspices of the International Biological Program (Bliss, 1977; Tieszen, 1978; Brown *et al.*, 1980; Bliss *et al.*, 1981) that has integrated analyses of both the abiotic and biotic environments from locations in the Low and High Arctic. Low light intensity, low air temperature, and low levels of soil nutrients limit plant growth (Chapin and Shaver, 1985), and these factors profoundly affect at least the purely physiological aspects of reproduction.

Precipitation is surprisingly low in the Arctic, and the northernmost portions are Polar Desert. Usually there is a major flush of moisture during the brief spring break-up when runoff from snow-melt is rapid and abundant. Soils can be saturated due to the very shallow depth of thaw and the impermeable, ice-rich permafrost only a short distance from the surface. Water that does not run off cannot infiltrate to depth and thus remains ponded in depressions. Yet water stress can be an important factor in the Low Arctic on well-drained topographic prominences of all sizes and throughout the more arid High Arctic.

FIG. 1 Ice-wedge polygon nets present a repeated pattern of dry, exposed rims and wet and often flooded troughs and centres over vast areas of coastal plains.

I shall never cease to marvel at the many forms of patterned ground that owe their existence to frequent and prolonged freezing temperatures in seasonally saturated, fine-grained soils (Fig. 1). Some of the major polygonal structures are products of previous colder intervals, but frost action is today responsible for both permanent and ephemeral microtopography that contribute importantly to a diversity of plant habitats. Physical disturbances of varying frequency, magnitude, and extent are very much a part of the arctic landscape, varying from mass-wasting on ridges and steep slopes to frost-churning in the fine-grained soils of depressions (Fig. 2). Frequent freeze–thaw cycles during the spring and

FIG. 2 A cluster of non-sorted circles or ''frost boils'' that effectively interrupt what would otherwise be a continuous sward of vegetation.

autumn, when soils tend to be moist, are the intervals of greatest disturbance, but even in summer needle-ice formation in bare soils (Fig. 3) creates conditions inimical to seedlings. Botanists working in the far north should go back to the early papers by Raup (1951), Sigafoos (1951, 1952), Hopkins and Sigafoos (1951, 1954) on the role of frost action as a determinant of vegetation patterns.

In addition to the forces of frost and gravity, the normal burrowing activities of the arctic ground squirrel and arctic fox are responsible for disturbances of consequence to the substrate (Price, 1971) and to plant diversity. Where grizzly bears have attempted to dig out ground squirrel burrows, the excavations and effects on plant life are even more marked. Lemmings and voles can also be destructive where they overwinter in large numbers. These disturbances, in the aggregate, increase the diversity of microsites and of plants (Fox, 1985a).

An ostensibly flat tundra plain is actually a complex of convexities and concavities: microhummocks a few centimetres high; raised beach ridges, stream banks, and lake shores of a metre or more (Fig. 4); or pingos several metres high. The uneven surface, with attendant extremes of soil moisture and surface instability, creates a mosaic of habitats on meso- and micro-scales. Add to this the larger geomorphic changes, such as those wrought by the erosion and deposition of river meanders, landslips on slopes, and thermokarst features (Figs 5 and 6), and this ''monotonous'' landscape is more accurately a myriad of plant habitats. *Spatial heterogeneity is extreme*, and it follows that patchiness of the plant cover can be marked.

FIG. 3 Caps of mineral soil are thrust upward with the growth of needle-ice and then returned to the surface as the crystals melt, leaving this characteristic texture to the soil.

FIG. 4 Stream banks, drained-lake shoreline, and earth hummocks provide topographic relief and sharply defined habitats over short distances on an otherwise featureless plain.

FIG. 5 Large ice mounds can form during one or two winters, persist for one year or perhaps a decade, and disappear in one or two years.

FIG. 6 Melting of ice-wedges breached the narrow divide between these two thaw-lakes suddenly allowing one to drain into the other. Lake shore and basin habitats are now available for colonization by plants.

The mesotopographic gradient, as expressed by Billings (1973), provides a model for integrating cold-climate physical processes with physiography and plant cover. Convexities are exposed to winds that dry the surface and even remove fine particles of soil where they are not protected by plants. This can mean a thin winter snow cover and therefore early exposure in the spring to diurnal cycles of freezing and thawing. The generally coarse, thin soil is low in organic content and poor in water-holding capacity. The brief flush of moisture from snow-melt and summer precipitation is rapidly drained away.

Slopes receive the colluvium from the summits, and the soils become progressively deeper and finer in texture downslope. Mass-wasting of the steeper slopes can preclude closed communities, and the plant cover may be limited to scattered plants with stout taproots or caespitose habit to compensate for the instability. Even slopes with a continuous plant cover can exhibit surface movement that slowly builds characteristic lobate and terraced landforms (Fig. 7). The products of mass-wasting accumulate in the basins, and the soils are deep, with abundant fines and a significant organic fraction. Drainage is poor, water is abundant, and the soils are saturated or flooded. Impeded drainage and deep fine-grained soils carry the liability of more intense frost action and permafrost close to the surface that is only partly ameliorated by a deeper, late-lying snow cover. Major irregularities can be catchments for snow (Fig. 8), which in turn controls the length of the growing season as well as the moisture regime of the snowbeds and wherever the meltwater flows.

An additional level of complexity is superimposed on the landscape by microtopography that forms a parallel series of topographic gradients; each small hummock has an exposed high point, a short slope, and a base. The summits and slopes are warmer and drier than the cooler, moister basins, although the total relief is only a few centimetres. Similarly, the microscale pattern produced by polygonal desiccation cracks exerts a strong influence on the structure of plant communities (therefore also on the genetic structure of populations; see Loveless and Hamrick, 1984) by providing warmer, moist, and sheltered sites for seedling survival where the physical environment otherwise precludes a plant cover (Sohlberg and Bliss, 1984).

On relatively level ground with only modest variations in relief there can be distinct shifts in the plant cover over very short horizontal distances. Changes in soil texture produce the same or similar effects on soil moisture as would be anticipated from changes in topography. The patterns of plants on floodplains and river terraces can be especially varied for this reason. Alluvium of different textures is present in complex patterns; coarse gravels lie adjacent to sands, and both can be capped by fine overbank deposits of varying thicknesses with consequent changes in drainage — the thicker the deposit of fines the more impeded the drainage.

FIG. 7 Gelifluction terraces create contrasting step and riser habitats that differ in the plant cover, soil moisture, and depth of seasonal thaw.

FIG. 8 Snowbanks that form in stream channels can last until late summer.

The plant cover is complex because the soils and terrain are irregular. The land is a

> . . . kaleidoscopic patchwork of disturbance intensities ranging from essential stability to activity in the soil so intense that no plants can survive in it. Further, this patchwork . . . changes with time, at rates that differ with slope, exposure, length of season, water supply and soil textures.
>
> (Raup, 1969)

The irregularities determine the environmental gradients. Where the landforms are abrupt, the gradients are steep, and the discontinuities in the vegetation are also likely to be similarly well marked. Therefore geomorphic history as well as contemporary processes determine the essential ingredients of spatial heterogeneity through variations in topography and soil texture that in turn determine slope, aspect, soil temperature, and soil moisture.

III. Arctic Flora and its Formation

What has shaped the flora and how have these species met the challenges of the arctic environment? The majority of the flora is composed of perennial herbs. There are only three or four arctic annuals, and I do not know that the number of species exhibiting biennial habit has been carefully assessed. The most common and widespread of the woody plants are the dwarf birches, willows, and several ericaceous shrubs. Not unexpectedly, the vascular plant families best represented (by species, not genera) are Asteraceae, Cyperaceae, Poaceae, Brassicaceae, Rosaceae, Caryophyllaceae, Salicaceae, Fabaceae, Saxifragaceae, and Scrophulariaceae.

Again and again it has been noted and emphasized that a significant proportion, in fact more than half, of the species are polyploid. The explanations have varied:

(1) the polyploid taxa are adaptively superior (hardy);
(2) the Arctic selects for characteristics associated with polyploidy, i.e. the high frequency is related to the high proportion of herbaceous perennials in the flora;
(3) polyploids excelled at migrating from late-glacial refugia to colonize postglacial landscapes;
(4) polyploids are buffered against genetic novelty thus genetic uniformity is assured, and this conservatism is selected for in the Arctic.

Stebbins (1984, 1985) has reviewed these hypotheses and drawn elements of each to form his secondary-contact hypothesis: stabilization by polyploidy of hybrids formed during post-glacial mingling of congeners, which until then had been isolated from one another by glaciers. Not only does secondary contact explain well the different frequencies of polyploids in floras at different latitudes, it is congruent with what we know about the past two million years or more of earth history.

The few data for the end of the Tertiary (Pliocene) indicate a coniferous woodland with floristic elements we recognize as part of modern tundra. From these boreal forests certain plants — those pre-adapted to the habitats destined to become more widespread during the Quaternary — remained after the forests themselves disappeared. To these were added waves of migrants to North America from mountains and steppes of central and northeastern Asia and from the mountains and plains to the south in North America. Thus a new floristic mix became established on the treeless landscape of the Arctic (Murray, 1981), a mix that has changed by additions and subtractions with each of several episodes of glacial expansion and contraction over a span of two million years (Murray *et al.*, 1983).

Beringia was crucial to this floristic development, for in this area between the Kolyma and Mackenzie rivers plants persisted, when to the east and west and south glaciers covered the land, floras were exchanged, and from there the plants

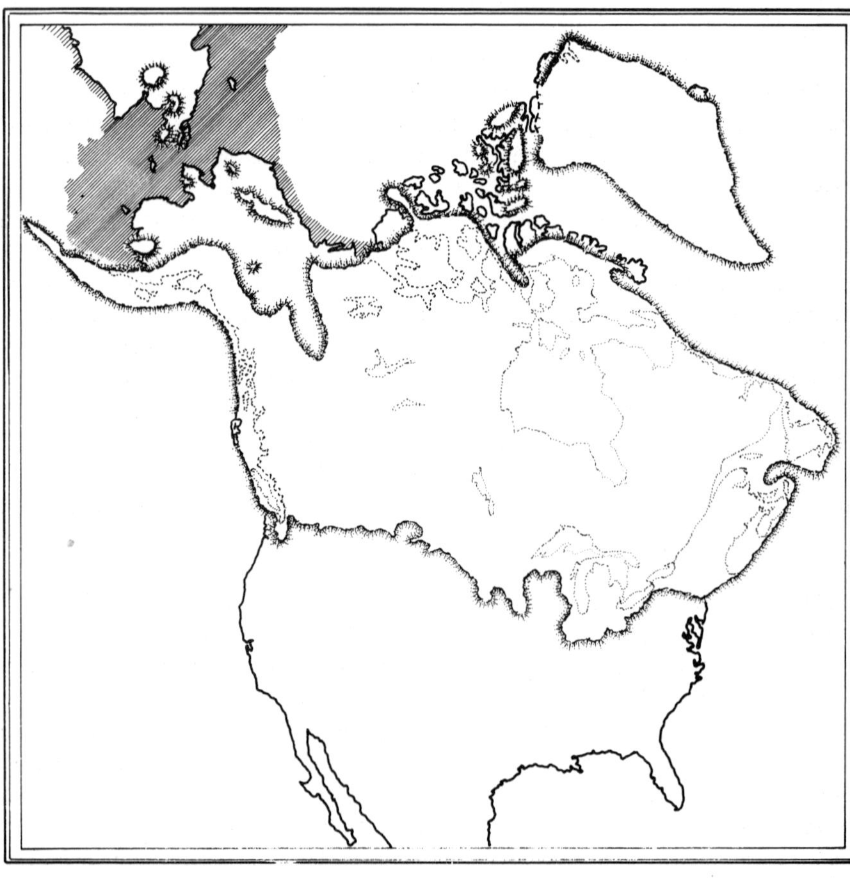

FIG. 9 Limits of ice sheets and mountain glaciation during the last glacial maximum. Shaded: the area of an emergent Bering Platform.

dispersed to reoccupy deglaciated surfaces. Figure 9 shows the approximate limits of ice sheets and mountain glaciation during the last glacial maximum (based on Dyke *et al.*, 1982; England, 1986; Funder, 1982; Hamilton and Thorson, 1983; Matthews, 1979; Porter *et al.*, 1983; and Velichko *et al.*, 1984). That the limits are approximate must be stressed, for the glacial history of the Arctic is currently being re-evaluated and hotly debated. Not all of the unglaciated areas shown could have supported a large flora due to presence of snowfields (Savile, 1961) and extreme aridity (Carter, 1981, 1983). Nevertheless, this recent work forces us to reconsider the number, size, and placement of refugia, especially in arctic Canada and Greenland. Note too, that the glacial margins were certainly punctuated by ice-free oceanic headlands and mountain peaks (collectively called *nunataks*) in the northeastern Gulf of Alaska in the west and from Baffin Island southward to the Gulf of St Lawrence in the east, as well as both west and east Greenland. Additional centres of refugial existence would only strengthen the secondary-contact hypothesis by providing more opportunities for fragmentation of populations, full-glacial allopatry, and subsequent opportunities for parapatry, sympatry, and hybridization during interglacial intervals.

Transberingian migrations, predominantly west to east, Asia to America, occurred during the peak of the glacial periods when the sea level was low enough to present an emergent Bering Platform (Fig. 9). Floristic elements were differentially favoured by the changing conditions as the land bridge formed and spread from an isthmus to its full continental connection and back again. At the same time, however, the heavily glaciated mountain ranges and massive ice sheets to the east isolated the refugial areas from the rest of North America. Conversely, during the interglacial periods that glacial barrier was removed, but the Bering Strait curtailed further contributions from Asia. Floristic exchanges shifted, therefore, to between eastern Beringia and populations from mountain refugia (nunataks) along the cordillera and from the plains south of the maximum extent of ice sheets. Although taiga quickly returned to fill the northern valleys as glaciers receded, tundra habitats probably persisted well into the Holocene along the interior of the western cordillera where a continental climate and abundant glacial outwash provided suitable conditions for the migration of tundra species along the mountain front.

Thus there were ample opportunities for the differentiation of populations in isolated, contracted areas followed by a release of constraints and expansion of ranges into new and demanding settings. Hybridization and formation of polyploids provided recombinants superior to the parental stock in the new environments. Habitats became progressively more heterogeneous as soils developed on glacial till and outwash and as the surface was reworked by frost action and by fluvial and slope-forming processes. The development of a continuous cover of vegetation in favourable sites altered energy exchange at the surface and allowed for the rise of permafrost in the soils and peat development concomitant with an acidification of the substrate. The sorting of the diploids and polyploids was presumably as described by Johnson and Packer (1965) at Ogotoruk Creek in northwestern Alaska: polyploids exploited in higher proportion the wet, cold sites with fine-grained soils subject to intense frost disturbance, the very sites that became more abundant or were more strongly modified by Quaternary events.

It would seem that numerous polyploids are *prima facie* evidence for a long history of outcrossing in this flora, and any trend away from sexual reproduction would have to be a recent one. If we can accept the essential instability and complexity of the physical and biological environments during any one interval of time and major changes in environments over time, then we must wonder how the plants have coped. Spatial unpredictability confronts the plant propagules, and the variability inherent in sexual reproduction would insure genetic potential for coping with that unpredictability. Asexual reproduction, merely a copying process, would not.

IV. Agamospermy

The conclusion of Löve (1959) that ". . . nowhere else is there a higher frequency of asexual seed formation" draws heavily upon Gustafsson (1947) and examples from so-called apomictic genera and the invocation of agamospermy to explain seed production in spite of irregular meiosis, observed or inferred from aborted pollen (Mulligan and Findley, 1970; Mosquin and Hayley, 1966).

Indeed, there are agamospermous plants in the Arctic, widespread and obviously successful taxa of *Antennaria, Calamagrostis, Potentilla*, and *Taraxacum* as well as *Rubus, Hieracium*, and *Alchemilla* (Greenland), genera that are notorious for their complex of microspecies. There are agamospermous species complexes in *Arnica* (Wolfe, 1980) and *Erigeron* (Spongeberg, 1971). It is also true that sexual diploid antecedents of agamospermous polyploids are present in the Arctic.

There is a clear relationship between agamospermy and polyploidy; nearly all gametophytic apomicts are polyploid, but not all polyploids are apomicts (de Wet, 1980). Agamospermy, especially in conjunction with herbaceous perennial habit, allows a stabilization of genotypes that if adaptive then permits rapid, local exploitation of new habitats. We should not consider the apomicts as "locked-in" to that mode nor consider a species apomictic throughout its range. Thinking back to how microspecies complexes form in the first place, we are reminded that apomictic groups are not obligatorily infertile. If we accept that polyploidy is a process commonly involving the fusion of unreduced and normal gametes (de Wet, 1980), the tetraploid being formed in two intermediate steps involving the formation of a triploid, then we have the potential to see triploids stabilized by agamospermy in genera without apomictic reputation but with polyploids.

Within *Taraxacum* there are sexual diploids, facultatively agamospermous triploids, partially sexual triploids, and tetraploids coexisting in varying frequencies (den Nijs and Sterk, 1980, 1984a,b; den Nijs, pers. comm.). The sexual diploids can be locally prominent. So much for the apomictic reputation of this genus.

It is also sobering to reflect on the case of *Smelowskia calycina*, which has a race with the chromosome number of $2n = 22$ derived from $2n = 12$ diploids by $n - 1$ aneuploidy superimposed on an intermediate polyploid increase. Although Greene (1978) suspected that seed production was by agamospermy, he was able to show that megasporogenesis was regular. How many other, similar suspicions of apomictic reproduction are also mistaken?

Seeds have become the subject of close scrutiny as the influence of plant population biology is felt in the study of tundra plants. To be sure, the constant renewal of plant cover in areas disturbed by natural and anthropogenic causes is accomplished by vegetative spread, but also by seeds from the buried seed pool and by normal dispersal of the annual seed production. In fact the production of seeds and their spread can be remarkable during good seed years. This closer look at the natural history of arctic plants has also revealed seedlings in high densities, although survival rates are low. It is therefore increasingly obvious that seeds are being produced in quantity, and it is neither correct nor necessary to invoke apomictic generation of them. Obviously generalizations about the efficacy of sexual reproduction in the Arctic have been incorrect.

If there is a shift of reproductive modes as one goes north, this certainly should be evident in the High Arctic. Beschel (1970) claimed

> a great decrease in sexual reproduction . . . across the boundary between arctic tundra and polar desert. The percentage of species which do not form seeds arising from pollination but propagate in various asexual ways doubles there, even when favourable years are considered.

This is countered by Savile (1972) who stated ". . . even in the high arctic, sexual reproduction remains the rule in flowering plants." As Bell and Bliss (1980) have pointed out, there are no novel strategies there beyond those already encountered in the Low Arctic, yet reproduction is even more limited by a shorter time for seeds to mature, and therefore fewer seeds are produced. Conditions are poorer for germination of seeds, and there is very low seedling survival. Surprisingly, Grulke and Bliss (1985) observed that vegetative reproduction decreases with latitude. Their analysis of the vascular flora at four localities in Alaska and Canada from 70 to 78°N shows the number of species reproducing vegetatively decrease dramatically. I note, however, that the percentage remains more or less the same.

Of seven species of *Saxifraga* studied by Grulke and Bliss (1985), six produced germinable seeds upon which four species relied for individual replacement. Three species relied on vegetative means, reliance in this case defined as having few flowers, low germinability of seeds, and a high carbon allocation to basal buds, bulbils, or stolons. Vegetative increases have not replaced seed production. They also reported what was to them an unexpectedly high allocation of carbon to flowering and fruiting, but explained it away as relictual — in my view an unjustified extrapolation from Mosquin (1966).

Not only are seeds formed, but they are also incorporated into the seed bank in large numbers. Freedman *et al.* (1982) analysed samples of the uppermost 2 cm of soil from several habitats, both disturbed and undisturbed, at 78°N and found the highest numbers of seeds in three disturbed sites (gravel pad, gravel beach, and fox den) where there were also the highest densities of seedlings. They concluded that sexual reproduction is most important to the herbaceous, ruderal species. On the other hand, seed banks in the Low Arctic and in mountain tundras to the south (Leck, 1980; Fox, 1983; McGraw, 1980; Roach, 1983; and Gartner *et al.*, 1983) provide evidence of buried seed pools for species of relatively stable vegetation types, involving the taxa prominent in forming closed communities.

V. Breeding Mechanisms

Of the plants obviously reproducing sexually, there is a question of the proportions employing cross-pollination or self-pollination. Lloyd (1980) observed that "self- and cross-fertilizing species are not equally common at all latitudes," and that many have "observed that the frequency of autogamy increases towards the north in the north temperate zone," and concluded that "at very high northern latitudes, cross-fertilizing species are uncommon." What, however, do we really know about breeding mechanisms from direct observations? Although there are numerous brightly coloured flowers in the Arctic, many of these with a marked aroma and obviously producing nectar (Swales, 1979), it has nevertheless been noted that the requisite insect pollinators are often conspicuous by their absence. Dipterans and bumblebees are the vectors most often implicated, but even when insects are present, the weather in the Arctic can be day after day too cold and blustery to permit flight. The northern limit of taxa adapted to pollination by bumblebees may in fact reflect the northern limit of the bee (Savile, 1959).

There are brightly coloured flowers that do not produce nectar, nectar-producers without obvious mechanisms for advertising their presence, and flowers morphologically adapted to insect visitors, yet without them. These paradoxes have been explained by Mosquin (1966), who, in a very challenging paper, put forward the hypothesis

> . . . that in the arctic, genetic uniformity is at a selective advantage. Therefore nearly every species has one or more reproductive devices for promoting population uniformity. Further, features of arctic plants which appear to promote variability, are, in fact, vestigial. Features such as showy flowers, nectar production, zygomorphic corollas, and perhaps genetic variability itself are not to be interpreted as evidence that outcrossing has adaptive function to-day but rather these features are evidence of outcrossing in ancestry.
>
> (Mosquin and Martin, 1967)

Thus, these outcrossing attributes are presumably the expression of the genetic baggage these plants carry as a result of their temperate zone ancestry where outcrossing was adaptive. The plants, according to Mosquin, are predominately selfers or apomicts. Since those papers were written, however, new data call into question such conclusions.

Enforced outcrossing is of course present in all dioecious taxa that are not also apomictic. Fox (1985b) even noted a relative increase in the incidence of dioecy northward, which in northern Alaska is accounted for by the ubiquitous presence of willows that remain in higher proportion than other taxa. To the work of Kevan (1972) at Camp Hazen on Ellesmere Island in the Canadian Arctic we can add the work of Tikhmenev (1984), whose studies on Wrangel Island and in the Okhotsk-Kolyma Uplands, USSR, have added to the list of obligate entomophiles several species of willows, all of them widespread. The list for Wrangel Island also includes the common bistort *Polygonum ellipticum* (a segregate of *P. bistorta* s. lat.) *Minuartia macrocarpa, Saxifraga oppositifolia, Dryas punctata* (a segregate of *D. octopetala* s. lat.), six species of legumes in two genera *Astragalus* and *Oxytropis, Castilleja elegans* and four of the very common *Pedicularis* species. I think it is just a matter of time, after more studies such as these, until the list will become even longer (Table 1).

In addition, Tikhmenev (1984) has distinguished two classes of self-compatible types: preferential entomophiles always producing far fewer seeds autogamously and facultative entomophiles producing near equal numbers of seeds by either outcrossing or selfing, and therefore essentially independent of insects for seeds. Kevan gave *Dryas integrifolia* and *Saxifraga oppositifolia* as preferential taxa to which Tikhmenev has added *Parrya nudicaulis, Claytonia arctica, Polemonium acutiflorum*, and *P. boreale*. The list of facultatively entomophilous species is of course larger and includes a broad range of genera from at least eight families. Obligate and preferential entomophily is reinforced by dichogamy (both protandry and protogyny) maintained for periods of several hours up to a few days.

There are divergent opinions on the classification of some taxa as obligate, preferential, or facultative entomophiles. That is not to say that someone is wrong, but that the behaviour of the species is not uniform throughout its range (e.g. *Saxifraga oppositifolia*, which is also chromosomally variable, and especially *Pedicularis*: see Kevan, 1972; MacInnes, 1972; Macior, 1975; Williams and Batzli, 1982; Tikhmenev, 1984). There are local, seasonal, and perhaps major regional differences in breeding behaviour, which will confound any attempt to generalize for the Arctic. What we learn from this is the complexity of reproductive options realized.

Kelso (1985, and in press) has reported a homostylous, self-compatible *Primula eximia* Greene as derivative from a distylous and presumably self-incompatible *P. tschuktschorum* Kjell. The homostylous species now occupies the larger range of the two and is locally very abundant, whereas the heterostylous species is a narrow endemic of the Bering Strait region. Yet this example from *Primula* does not allow a generalization in favour of autogamy, for Kelso noted that three other heterostylous species occur and are successful in the same environment. Insect attractants have been retained in the monomorphic self-compatible populations,

TABLE 1
Entomophilous pollination in arctic species

Taxon	Obligate	Preferential	Facultative
Andromeda polifolia			T
Androsace ochotensis (*Douglasia ochotensis*)	T		
Arctous alpina (*Arctostaphylos alpina*)		T	
Astragalus alpinus	T		
Astragalus richardsonii (*A. aboriginum*)	T		
Astragalus umbellatus	T		
Claytonia acutifolia			T
Claytonia arctica		T	
Caltha arctica			T
Cassiope ericoides			T
Cassiope tetragona			K
Castilleja elegans	T		
Diapensia obovata		T	
Draba cinerea			T
Draba hirta			T
Dryas integrifolia	K		
Dryas punctata (*D. octopetala*)	T		
Empetrum nigrum			T
Epilobium angustifolium			T
Epilobium latifolium			T,K?
Ledum decumbens			T
Lesquerella arctica		K?	
Loiseleuria procumbens			T
Minuartia arctica			T
Minuartia macrocarpa	T		
Minuartia rubella			T
Oxytropis adamsiana	T		
Oxytropis czukotica	T		
Oxytropis evenorum	T		
Oxytropis gorodkovii (*O. nigrescens* ssp. *pygmaea*)	T		
Oxytropis wrangelii	T		
Parrya nudicaulis	T		
Pedicularis adamsii	T		
Pedicularis amoena	T		
Pedicularis capitata	T,K	W	
Pedicularis lanata		W	
Pedicularis langsdorffii (incl. *P. arctica*)	T	K,W	

K, Kevan (1972), Ellesmere Island, NWT, Canada.
T, Tikhmenev (1984), Wrangel Island and Okhotsk-Kolyma Upland, USSR.
W, Williams and Batzli (1982), Atqasuk, Alaska, U.S.A.

Taxon	Obligate	Preferential	Facultative
Pedicularis lapponica	T	W	
Pedicularis oederi	T		
Pedicularis sudetica	T	K	W
Pedicularis verticillata	T	T	
Polemonium acutiflorum		T	
Polemonium boreale		T	
Polygonum ellipticum (*P. bistorta*)	T		
Polygonum ochotense	T		
Potentilla elegans	T		
Potentilla nivea			T
Potentilla uniflora			T
Phyllodoce coerulea			T
Ranunculus pygmaeus			T
Ranunculus sulphureus			T
Rhododendron aureum		T	
Rhododendron parviflorum			T
Rhododendron redowskianum			T
Salix arctica	K		
Salix fuscescens	T		
Salix glauca	T		
Salix phlebophylla	T		
Salix pulchra	T		
Salix reptans	T		
Salix reticulata	T		
Salix sphenophylla	T		
Salix tschuktschorum	T		
Saxifraga hirculus			T
Saxifraga nelsoniana			T
Saxifraga oppositifolia	T	K	
Saxifraga redowskiana			T
Saxifraga tricuspidata			K
Senecio atropurpureus			T
Senecio frigidus			T
Sieversia pusilla	T		
Spiraea stevenii (*S. beauverdiana*)			T
Stellaria ciliatosepala			T
Stellaria fischerianum			T
Stellaria jacutica			T
Taraxacum arctogenum			K?
Vaccinium uliginosum			T
Vaccinium vitis-idaea			T

K, Kevan (1972), Ellesmere Island, NWT, Canada.
T, Tikhmenev (1984), Wrangel Island and Okhotsk-Kolyma Upland, USSR.
W, Williams and Batzli (1982), Atqasuk, Alaska, U.S.A.

and early dichogamy temporarily sustains an outcrossing mode (see also Campbell *et al.*, 1986, re *P. laurentiana*). Self-compatibility *per se* does not prevent cross-pollination from occurring (see also Baker, 1966, re *Armeria*).

There have been similar shifts in heterostylous *Armeria* and *Menyanthes* in which the arctic and disjunct alpine populations are homostylous: their establishment and their continuing requirement to persist in small, isolated populations following long distance dispersal was favoured by self-compatibility (Baker, 1959). Entomophily (and presumably also anemophily) is functional in arctic plants, thus the assumption that morphological features and energetic investments for outcrossing are vestigial cannot be sustained. That many of these outcrossing species are self-compatible does not signal a shift from one mode to another.

VI. Ecological Differentiation

Additional insights into the nature of species (and therefore also of the processes forming them) come from the botanists who are faced with the task of defining plant associations. They must deal with a flora of the same widespread species growing on demonstrably different sites. This versatile behaviour of species forces one to recognize communities on the basis not of indicator species, but in terms of relative quantity or to describe the plant cover of habitats and landforms. The different landforms are discontinuous units, but the overlap in species composition of the vegetation on them is a problem and a clue to the makeup of those species. Their versatility derives either from a single genotype that confers physiological plasticity or multiple genotypes for any one morphological species.

Raup (1969) carefully analysed the behaviour of species with respect to gradients of total cover, soil moisture, and physical disturbance at Mesters Vig in northeast Greenland. His analysis is based upon 157 species observed at 300 sites plus the wisdom gained from a lifetime of studying the arctic flora. Almost half of the species showed a wide tolerance to variations in plant cover; they were found isolated on otherwise bare ground as well as on tundra covered at medium to high densities. Therefore, he concluded that these were essentially independent of communal relationships, hence, the floristic similarity of different communities. Only 14% had a low tolerance to variations in cover and the potential of being indicator species, yet many of these were rare. A large proportion of the species had a wide tolerance to variation in soil moisture; 70% of them occurred in two-thirds of the gradient. Again, those preferentially segregated on the moisture gradient were rare to occasional, or if common were in the moist sites. Almost one-third of the species had a wide tolerance to disturbance. One-half had showed a narrow tolerance to variations in disturbance, but nearly all of these were on the stable soils.

In terms of general frequency, Raup found that two-thirds of the flora was common to abundant and one-third was generally occasional to rare. The widely tolerant species were far more likely to be among the common to abundant taxa than those narrowly tolerant; the narrowly tolerant species were much more abundant among the rare and locally occasional species. Raup concluded that there are gradients of response within species interacting with factor gradients on the landscape. The variants in species are arrayed selectively according to changes in site factors. The species have a built-in flexibility, and I take that to mean they are ecotypically differentiated and richly so.

Some of these ecotypes may be reproductively conservative, but McGraw and Antonovics (1983) have described how plant–site interactions in *Dryas octopetala* growing on fellfields and adjacent, essentially contiguous, snowbeds have resulted in distinctly adapted populations, notwithstanding insect pollination and an absence of genetic barriers to crossing between ecotypes. The strong selection observed against seedlings of one ecotype in the site of the other caused them to predict a steep gradient in gene frequencies between the two habitats. The respective coherence of the two populations was not due to this selection alone, for the phenological behaviour of the plants and the foraging behaviour of the insects led to exchanges of pollen almost totally among flowers of the same ecotype.

Tolerances of a species to factor gradients are not the same from place to place throughout the Arctic. Behaviour at Mesters Vig cannot be extrapolated to a similar site in the Canadian Arctic or to northern Alaska. It is reasonable, therefore, to presume that widespread species, the circumpolar element of roughly 600 species for example, are differentiated in ways certainly physiological and often accompanied by morphological change; both changes presumably adaptative to local factors. I suggest that the edaphic differences associated with meso- and microtopography provide sufficiently strong contrasts for differentiation to occur. Thus, species that appear extremely tolerant most likely consist of populations of ecotypes or local races. Some of the tolerant species could be genomically differentiated as well, for the more we count the chromosomes of our Alaskan plants, the more we find multiple numbers for the same morphological species.

VII. Conclusions

Generalizations help organize data, but their truth can be diminished by a failure to acknowledge complexity. To characterize the Arctic as uniformly harsh only reflects our temperate zone backgrounds. The predictable harshness attributed to the Arctic oversimplifies the limitations plants face in that environment. These are generally seen as minimum values for temperature, moisture, nutrients, etc. that test the limits or exceed the physiological tolerance of plants. This bias makes it difficult not to view plants in the Arctic as always perilously close to the very limits of life (Raup, 1969). Raup mused

> what we need is a first class Eskimo botanist — one who thinks of the tundra as home, and a very good place to live. I think he would see the plants as they are, members of an ancient flora remarkably well adjusted to the habitat.

The Arctic is not spatially uniform at any scale. It is climatically variable from year to year, yet predictable within limits. Trends of change are detectable over centuries and there is ample evidence of major climatic changes over millennia. The flora, with a history that exceeds two million years, developed through periods of contrasting demands imposed by the physical and biological environments. Cycles of change have involved the entire region, flora, and fauna, affecting not only the stage, but all the players as well. Therefore, I believe Quaternary history has been one of the most important factors in the evolution of the arctic flora, having determined: (1) varied extraterritorial influxes; (2) extirpation, displacement, or fragmentation of populations on a broad geographic scale; (3) dynamic secondary contact as isolated populations expand their ranges until contiguous with congeners; and (4) landform and substrate.

The substitution of sexual reproduction by asexual means in the Arctic is incorrect, no matter how well that belief is embedded in the literature. Asexual reproduction is successful and therefore important in numerous instances, but only rarely is it exclusive. The conclusion that autogamy is prevalent is similarly unsubstantiated. Self-compatibility and cross-pollination are not mutually exclusive, thus I suggest that mixed mating is common and highly advantageous, since it allows for success at two levels: (1) the virtual guarantee of reproduction in bad years for pollinators; and (2) both the origin and maintenance of variation as well as the coherence of adaptive elements.

There are hazards in generalizing behaviour for a species or a region as if all populations of the species or all taxa for the region could be expected to behave alike. This adaptationist view is confounded by phyletic constraints, even stasis. Paradigms must take these realities into account: the tendency of buttercups to behave as buttercups.

The question of breeding systems in the Arctic should be re-examined, for we now have the means to measure and to assess the effects of one mode or the other on the genetic structure of populations. The Arctic offers not the simple system one often hears about, but one complicated by dioecious anemophilous and entomophilous species, gynodioecious taxa (e.g. *Stellaria longipes* s. lat.: Philipp, 1972, 1975), self-incompatible diploids, self-compatible polyploids, and apomictic polyploids. Vegetative capacity is well developed among many of the herbaceous perennials and is clearly of importance. The means exist for the majority of the plants to produce seeds, apomictically in some instances, but sexually in the majority of others. How often seed production occurs — annually, every few years, or infrequently — is another question. The results projected from mixed mating depend on the proportions of self- and cross-pollination. Whichever the case, the theoretical outcomes are affected by conservatism, a function of perennial habit, long generation time, and replacement of individuals in part by buried seed pools of undetermined ages.

Acknowledgements

My thanks to W. S. Armbruster, S. Kelso, B. M. Murray, H. M. Raup, D. B. O. Savile, and K. Schwaegerle for their critiques of the manuscript.

References

Alexandrova, V. D. (1977). Geobotanical Subdivisions of the Arctic and Antarctic. Komarov Lecture 29. Komarov Bot. Inst. Acad. Sci., Leningrad. In Russian; translated by Doris Löve (1980) as "The Arctic and Antarctic: Their Division Into Geobotanical Areas". Cambridge University Press, Cambridge.

Baker, H. C. (1959). The contribution of autecological and genecological studies to our knowledge of the past migrations of plants. *Am. Nat.* **871**, 255–272.

Baker, H. C. (1966). The evolution, functioning and breakdown of heteromorphic incompatibility systems. I. The Plumbaginaceae. *Evolution* **20**, 349–368.

Bell, K. L. and Bliss, L. C. (1980). Plant reproduction in a high arctic evironment. *Arct. Alp. Res.* **12**, 1–10.

Beschel, R. E. (1970). The diversity of tundra vegetation. *In* "Productivity and Conservation in Circumpolar Lands" (W. A. Fuller and P. G. Kevan, eds), pp. 85–92. Intern. Union Conserv. Nat. and Nat. Res. Publ., New Ser. 16.

Billings, W. D. (1973). Arctic and alpine vegetations: similarities, differences, and susceptibility to disturbance. *Bioscience* **23**, 697–704.

Billings, W. D. (1974). Arctic and Alpine vegetation: plant adaptations to cold summer climates. *In* "Arctic and Alpine Environments" (J. D. Ives and R. G. Barry, eds), pp. 403–443. Methuen, London.

Billings, W. D. and Mooney, H. A. (1968). The ecology of arctic and alpine plants. *Biol. Rev.* **43**, 481–529.

Bliss, L. C. (1971). Arctic and alpine life cycles. *Ann. Rev. Ecol. Syst.* **2**, 405–438.

Bliss, L. C. (ed.) (1977). "Truelove Lowland, Devon Island, Canada: a High Arctic Ecosystem". University of Alberta Press, Edmonton.

Bliss, L. C. (1981). North American and Scandinavian tundras and polar deserts. *In* "Tundra Ecosystems: a Comparative Analysis" (L. C. Bliss, O. W. Heal and J. J. Moore, eds), pp. 8–24. Cambridge University Press, Cambridge.

Bliss, L. C., Heal, O. W. and Moore, J. J. (eds) (1981). "Tundra Ecosystems: a Comparative Analysis". Cambridge University Press, Cambridge.

Brown, J., Miller, P. C., Tieszen, L. L. and Bunnell, F. L. (eds) (1980). "An Arctic Ecosystem: the Coastal Tundra at Barrow, Alaska". Dowden, Hutchinson and Ross, Stroudsburg.

Campbell, C. S., Famous, N. C. and Zuck, M. G. (1986). Pollination biology of *Primula laurentiana* (Primulaceae) in Maine. *Rhodora* **88**, 253–260.

Carter, L. D. (1981). A Pleistocene sand sea on the Alaskan Arctic Coastal Plain. *Science* **211**, 381–383.

Carter, L. D. (1983). Fossil sand wedges on the Alaskan Coastal Plain and their paleoenvironmental significance. *In* "Permafrost: Fourth International Conference Proceedings", pp. 109–114. National Academy Press, Washington.

Chapin, F. S. III and Shaver, G. R. (1985). Arctic. *In* "Physiological Ecology of North American Plant Communities" (B. F. Chabot and H. A. Mooney, eds), pp. 16–40. Chapman and Hall, New York.

Chapin, F. S. III, Tieszen, L. L., Lewis, M. C., Miller, P. C. and McCown, B. H. (1980). Control of tundra plant allocation patterns and growth. *In* "An Arctic Ecosystem: the

Coastal Tundra at Barrow, Alaska'' (J. Brown, P. C. Miller, L. L. Tieszen and F. L. Bunnell, eds), pp. 140–185. Dowden, Hutchinson and Ross, Stroudsburg.

Dyke, A. S., Dredge, L. A. and Vincent, L. (1982). Configuration and dynamics of the Laurentide ice sheet during the Late Wisconsin maximum. *Geogr. Phys. Quatern.* **35**, 5–14.

England, J. (1986). The last great mythological ice sheets. *Abstr. Am. Quatern. Ass.* 9th Annual Meeting, pp. 20–22. Champaign, Urbana.

Fox, J. F. (1983). Germinable seed banks of interior Alaskan tundra. *Arct. Alp. Res.* **15**, 405–411.

Fox, J. F. (1985a). Plant diversity in relation to plant reproduction and disturbance by voles in Alaskan tundra communities. *Arct. Alp. Res.* **17**, 199–204.

Fox, J. F. (1985b). Incidence of dioecy in relation to growth form, pollination and dispersal. *Oecologia* **67**, 244–249.

Freedman, B., Hill, N., Svoboda, J. and Henry, G. (1982). Seed banks and seedling occurrence in a high arctic oasis at Alexander Fjord, Ellesmere Island, Canada. *Can. J. Bot.* **60**, 2112–2118.

Funder, S. (1982). Planterefugiernei Grønland. *Naturens Verden* **7**, 241–255.

Gartner, B. L., Chapin, F. S. III and Shaver, G. R. (1983). Demographic patterns of seedling establishment and growth of native graminoids in an Alaskan tundra disturbance. *J. Appl. Ecol.* **20**, 965–980.

Greene, C. W. (1978). A Nomarski interference study of megasporogenesis and megagametogenesis in *Smelowskia calycina* (Cruciferae). *Am. J. Bot.* **65**, 353–358.

Grulke, N. E. and Bliss, L. C. (1985). Growth forms, carbon allocation, and reproductive patterns of high arctic saxifrages. *Arct. Alp. Res.* **17**, 241–250.

Gustafsson, A. (1947). Apomixis in higher plants. Part III: Biotype and species formation. *Lunds Univ. Arsskr.* **43**, 183–370.

Hamilton, T. D. and Thorson, R. M. (1983). The cordilleran ice sheet in Alaska. *In* ''Late Quaternary Environments of the United States'', Vol. 1, The Late Pleistocene (S. C. Porter, ed.), pp. 38–52. University of Minnesota Press, Minneapolis.

Hare, F. K. and Ritchie, J. (1972). The boreal bioclimates. *Geogr. Rev.* **62**, 333–365.

Hopkins, D. M. and Sigafoos, R. S. (1951). Frost action and vegetation patterns on Seward Peninsula, Alaska. *Geol. Surv. Bull.* 974-C.

Hopkins, D. M. and Sigafoos, R. S. (1954). Role of frost thrusting in the formation of tussocks. *Am. J. Sci.* **252**, 55–59.

Johnson, A. W. and Packer, J. G. (1965). Polyploidy and environment in arctic Alaska. *Science* **148**, 237–239.

Johnson, P. L. (1969). Arctic plants, ecosystems and strategies. *Arctic* **22**, 341–355.

Johnson, P. L. and Tieszen, L. L. (1973). Vegetative research in arctic Alaska. *In* ''Alaskan Arctic Tundra'' (M. E. Britton, ed.), pp. 169–198. Arctic Institute of North America Technical Paper No. 25.

Kelso, S. (1985). Heterostyly and homostyly in two arctic species of *Primula*. Abstracts of the 3rd International Congress on Systems Evolutionary Biology, Brighton, U.K.

Kelso, S. (1987). *Primula tschuktchorum* and *Primula eximia* (Sect. *Crystalophlomis*): a distylous species and its homostylous derivative from the Bering Sea region, Alaska. *Brittonia* **39**, 63–72.

Kevan, P. G. (1972). Insect pollination of high arctic flowers. *J. Ecol.* **60**, 831–847.

Leck, M. A. (1980). Germination in Barrow, Alaska, tundra soil cores. *Arct. Alp. Res.* **12**, 343–349.

Lloyd, D. G. (1980). Demographic factors and mating patterns in Angiosperms. *In* ''Demography and Evolution in Plant Populations'' (O. T. Solbrig, ed.), pp. 67–88. *Bot. Monogr.* **15**, University of California Press, Berkeley.

Löve, A. (1959). Problems of the Pleistocene and Arctic. *Publ. McGill Univ. Mus.* **1**, 82–95.

Loveless, M. D. and Hamrick, J. L. (1984). Ecological determinants of genetic structure in plant populations. *Ann. Rev. Ecol. Syst.* **15**, 65–95.

MacInnes, K. L. (1972). Reproduction in arctic populations of *Pedicularis* (Scrophulariaceae). Ph.D. Thesis, Department of Biology, University of Western Ontario, London.

Macior, L. W. (1975). The pollination ecology of *Pedicularis* (Scrophulariaceae) in the Yukon Territory. *Am. J. Bot.* **62**, 1065–1072.

Matthews, J. V. (1979). Tertiary and Quaternary environments: historical background for an analysis of the Canadian insect fauna. *In* "Canada and Its Insect Fauna" (H. V. Danks, ed.), pp. 31–86. *Entomol. Soc. Can. Mem.* **108**.

McGraw, J. B. (1980). Seed bank size and distribution of seeds in cottongrass tussock tundras, Eagle Creek, Alaska. *Can. J. Bot.* **58**, 1607–1611.

McGraw, J. B. and Antonovics, J. (1983). Experimental ecology of *Dryas octopetala* ecotypes. I. Ecotypic differentiation and life-cycle stages of selection. *J. Ecol.* **71**, 879–897.

Mosquin, T. (1966). Reproductive specialization as a factor in the evolution of the Canadian flora. *In* "The Evolution of Canada's Flora" (R. L. Taylor and R. A. Ludwig, eds), pp. 43–65. Toronto University Press, Toronto.

Mosquin, T. and Hayley, D. E. (1966). Chromosome numbers and taxonomy of some Canadian arctic plants. *Can. J. Bot.* **44**, 1209–1218.

Mosquin, T. and Martin, J. E. H. (1967). Observations on the pollination biology of plants on Melville Island, N. W. T., Canada. *Can. Field. Nat.* **81**, 201–205.

Mulligan, G. A. and Findlay, J. N. (1970). Sexual reproduction and agamospermy in the genus *Draba*. *Can. J. Bot.* **48**, 269–270.

Murray, D. F. (1981). The role of arctic refugia in the evolution of the arctic vascular flora — a Beringian perspective. *In* "Evolution Today, Proceedings of the 2nd International Congress of Systematics and Evolutionary Biology" (G. G. E. Scudder and J. L. Reveal, eds), pp. 11–20. Hunt Institute of Botany Document, Pittsburgh, PA.

Murray, D. F., Murray, B. M., Yurtsev, B. A. and Howenstein, R. (1983). Biogeographic significance of steppe vegetation in subarctic Alaska. *In* "Permafrost: Fourth International Conference Proceedings", pp. 883–888. National Academy Press, Washington.

den Nijs, J. C. M. and Sterk, A. A. (1980). Cytogeographical studies of *Taraxacum* sect. *Taraxacum* (= sect. *Vulgaria*) in Central Europe. *Bot. Jahrb. Syst.* **101**, 527–554.

den Nijs, J. C. M. and Sterk, A. A. (1984a). Cytogeography of *Taraxacum* sect. *Taraxacum* and sect. *Alpestris* in France and adjacent parts of Italy and Switzerland, including some taxonomic remarks. *Acta Bot. Neerl.* **33**, 1–34.

den Nijs, J. C. M. and Sterk, A. A. (1984b). Cytogeography and cytotaxonomy of some *Taraxacum* sections in Belgium and northern France. *Acta Bot. Neerl.* **33**, 431–455.

Philipp, M. (1972). The *Stellaria longipes* group in N. W. Greenland. Cytological and morphological investigations. *Bot. Tidsskr.* **67**, 64–75.

Philipp, M. (1975). Flower biology of *Stellaria longipes*. *Bot. Tidsskr.* **69**, 239–244.

Polunin, N. (1951). The real Arctic: suggestions for its delimitation, subdivision, and characterization. *J. Ecol.* **39**, 308–315.

Porter, S. C., Pierce, K. L. and Hamilton, T. D. (1983). Late Wisconsin mountain glaciation in the western United States. *In* "Late-Quaternary Environments of the United States", Vol. I, The Late Pleistocene (S. C. Porter, ed.), pp. 71–111. University of Minnesota, Minneapolis.

Price, L. W. (1971). Geomorphic effect of the arctic ground squirrel in an alpine environment. *Geogr. Ann.* **53**, 100–106.

Raup, H. M. (1951). Vegetation and cryoplanation. *Ohio J. Sci.* **51**, 105–116.
Raup, H. M. (1969). Studies of vegetation in northwest Greenland. *In* "Essays in Plant Geography and Ecology" (K. N. H. Greenidge, eds), pp. 45–62. Nova Scotia Museum, Halifax.
Roach, D. A. (1983). Buried seed and standing vegetation in two adjacent tundra habitats, northern Alaska. *Oecologia* **60**, 359–364.
Savile, D. B. O. (1959). The botany of Somerset Island, District of Franklin. *Can. J. Bot.* **37**, 959–1002.
Savile, D. B. O. (1961). The botany of the northwestern Queen Elizabeth Islands. *Can. J. Bot.* **39**, 909–942.
Savile, D. B. O. (1972). Arctic adaptations in plants. Monograph 6, Canadian Department of Agriculture, Ottawa.
Sigafoos, R. S. (1951). Soil instability in tundra vegetation. *Ohio J. Sci.* **51**, 281–298.
Sigafoos, R. S. (1952). Frost action as a primary physical factor in tundra plant communities. *Ecology* **33**, 480–487.
Sohlberg, E. H. and Bliss, L. C. (1984). Microscale pattern of vascular plant distribution in two high arctic plant communities. *Can. J. Bot.* **62**, 2033–2042.
Spongererg, S. A. (1971). A systematic and evolutionary study of North American arctic and alpine monocephalous species of *Erigeron* (Compositae). Ph.D. Thesis, University of North Carolina, Chapel Hill.
Stebbins, G. L. (1984). Polyploidy and distribution of arctic-alpine flora: new evidence and a new approach. *Bot. Helv.* **94**, 1–13.
Stebbins, G. L. (1985). Polyploidy, hybridization, and the invasion of new habitats. *Ann. Mo. Bot. Gard.* **72**, 824–832.
Swales, D. E. (1979). Nectaries of certain arctic and sub-arctic plants with notes on pollination. *Rhodora* **81**, 363–407.
Tieszen, L. L. (ed.) (1978). "Vegetation and Production Ecology of an Alaskan Arctic Tundra". Springer, New York.
Tikhmenev, E. A. (1984). Pollination and self-pollinating potential of entomophilic plants in arctic and mountain tundras of the northeastern USSR. *Soviet J. Ecol.* **15**, 166–172. 1985 translation from the Russian original, *Ekologiya* **4**, 8–15.
Velichko, A. A., Isayeva, L. L., Makeyev, V. M., Matishov, G. G. and Faustova, M. A. (1984). Late Pleistocene glaciation of the arctic shelf, and reconstruction of Eurasian ice sheets. *In* "Late Quaternary Environments of the Soviet Union" (A. A. Velichko, ed.), pp. 35–41. University of Minnesota Press, Minneapolis.
de Wet, J. M. J. (1980). Origins of polyploids. *In* "Polyploidy: Biological Relevance" (W. H. Lewis, ed.), pp. 3–15. Plenum, New York.
Williams, J. B. and Batzli, G. O. (1982). Pollination and dispersion of five species of lousewort (*Pedicularis*) near Atkasook, Alaska, USA. *Arct. Alp. Res.* **14**, 59–74.
Wolfe, S. J. (1980). Cytogeographical studies in the genus *Arnica* (Compositae: Senecioneae). 1. *Am. J. Bot.* **67**, 300–308.
Young, S. B. (1971). The vascular flora of St. Lawrence Island with special reference to floristic zonation in the arctic regions. *Contrib. Gray Herb.* **201**, 11–115.
Yurtsev, B. A., Tolmatchev, A. I. and Rebristaya, O. V. (1978). The floristic delimitation and subdivision of the Arctic. *In* "The Arctic Floristic Region" (B. A. Yurtsev, ed.), pp. 9–104. Nauka, Leningrad (in Russian).

Index